Environmental Problem Solving

Environmental Problem Solving

A HOW-TO GUIDE

Jeffrey W. Hughes

University of Vermont Press
Burlington, Vermont

Published by University Press of New England
Hanover and London

University of Vermont Press
Published by University Press of New England,
One Court Street, Lebanon, NH 03766
www.upne.com
© 2007 by University of Vermont Press
Printed in the United States of America
5 4 3 2 1

Library of Congress Cataloging-in-Publication Data

Hughes, Jeffrey W.
Environmental problem solving : a how-to guide / Jeffrey W. Hughes.
    p.   cm.
Includes bibliographical references and index.
ISBN-13: 978–1–58465–592–3 (pbk. : alk. paper)
ISBN-10: 1–58465–592–5 (pbk. : alk. paper)
1. Environmental management.   2. Problem solving.   I. Title.
GE300.H83 2007
363.705—dc22   2006102083

University Press of New England is a member of the Green Press Initiative. The paper used in this book meets their minimum requirement for recycled paper

To Lisa, Sam, Julia and all others who try to make the world a better place

Contents

Preface

It's easy to complain about the status quo or about the world falling down around your ears. It's less easy to do something about it. *Environmental Problem Solving: A How-To Guide* is for those who want to make things better but don't know how.

Most people who work on environmental problems eventually come to realize that passion, brainpower, and great intentions alone aren't enough to solve environmental problems. That realization can be devastating. After putting everything you have into a cause and seeing that your efforts haven't amounted to a hill of beans, why bother even trying?

Too many agency professionals, environmentalists, consultants, naturalists, interns, volunteers, environmental scientists, and concerned citizens fall prey to this level of discouragement, and that's what initially prompted me to write the book: to help professionals and lay people believe and demonstrate that they *can* solve environmental problems effectively and efficiently.

My original thinking (to produce a practical, user-friendly, how-to guide of environmental problem-solving tools and techniques) hasn't changed, but I have expanded the intended audience to include high school and college students and educators. On advice from faculty colleagues, the style of writing was adjusted a bit to make it more palatable to students.

If you've read this far in the preface, it's probably because someone made you do so—who but the author would ever willingly read a preface? The good news is that you are to be rewarded for your steadfastness. Flip to the last seven lines of page 6. Those seven lines tell you half of what you need to know about environmental problem solving.

(The other half of what you need to know is revealed in the pages that precede and follow those seven lines).

Acknowledgments

Among those I'd most like to thank are the people who sacrifice their evenings and weekends to serve on planning boards, conservation commissions, and other citizen groups. Their selfless efforts to make the world a better place (accompanied by their impassioned pleas for help) motivated me to start writing this book.

Hundreds of students in my undergraduate course, "Environmental Problem Solving and Impact Assessment," alerted me to shortcomings in earlier drafts of the book. Thank you, NR206ers, for your befuddled looks and blank stares—those reactions pushed me to make the text more understandable.

Past and present graduate students in the Field Naturalist and Ecological Planning programs at the University of Vermont provided invaluable reviews, for all had a realistic understanding of what was needed by professional practitioners in the real world. Other reviewers whose suggestions helped me improve the manuscript immensely were Dr. Richard Bowden, Dr. Gene Willeke, Dr. Toby Fulwiler, Dr. John Donnelly, Dr. Clifford Ochs, Dr. Hub Vogelmann, Wendy Cass, Matt Kolan, Abby Hood, Thom McEvoy, Lisa Meyer, Walter Poleman, and several anonymous reviewers. Virginia Hughes edited the manuscript before submission and Lillian ("Porky") Reade formatted the entire manuscript mess to make it more intelligible. John Robison, author, artist, and field naturalist, created the polluted river and DOC's KEY sketches. Dr. Phyllis Deutsch (Editor in Chief at University Press of New England) and Dr. Deane Wang (Associate Dean, University of Vermont) kept me from abandoning ship and plastering the manuscript on the web.

The University of Vermont granted me a sabbatical leave to work on this book, and the Conservation and Research Foundation provided financial support when I needed it most. Thank you to both.

I also thank Terry Sharik, Dean of the School of Natural Resources at Utah State University, for hosting me as I worked through the first draft. In addition to giving thoughtful feedback to early and late drafts of the manuscript, Dean Sharik worked tirelessly to sustain my sense of self-worth by letting me catch more and bigger trout than he did. Thank you, Dean Sharik.

Last, I thank Lisa, Julia, Sam, Stuart, Richard, Susan, Ginny, and Barbara for making every day a rich, worthwhile adventure. Thank you!

Example of a typical environmental problem—a confusing mish-mash of many problems.

Define
problem

Objectives

Constraints

How to solve environmental problems efficiently and effectively—"DOC'S KEY."

Environmental Problem Solving

What Problem-Solving Success (and Failure) Looks Like

Unfortunately, it's all too easy to find examples of well-intentioned problem-solving efforts that went astray. It's not so easy to find entirely successful problem-solving efforts. You're going to change that.

On the lines below, describe an environmental problem that you'd really like to see solved:

_____

_____

_____

Good. Hold onto this page, don't lose it. By the end of this book, *if you read carefully and thoughtfully and put real effort into the practice exercises,* you will have made major headway in solving at least part of this problem. That's a promise.

What Smart People Don't Understand (but You Will)

There are many, many reasons why smart people so often fail in their attempts to solve environmental problems. To reveal a few of them, try your hand at solving the following problem:

On public lands in the arid West, overgrazing by livestock has turned thousands of acres of once-productive rangeland into barren wasteland. Streamside (riparian) zones passing through these landscapes continue to produce abundant, high quality forage, but it is heavily grazed by livestock. This is of concern because intact riparian zones protect water quality and serve as hotspots for biodiversity. The conflict between environmentalists and ranchers over use of public lands in the arid West has become extreme.

Attack this problem as an impassioned environmentalist: what would you propose as a good solution? (Write your solution on the lines below.)

_____

_____

_____

Attack this same problem as a second-generation rancher: what would you propose as a good solution? (Write your solution on the lines below.)

_____

_____

_____

How do your solutions compare with those offered by others? Here's the type of solution that many impassioned environmentalists might propose:

"In return for not totally revoking the privilege of ranchers to graze on public lands, grazing fees for ranchers should be quadrupled, livestock should be fenced out of all streamside areas, less public land should be made available for livestock in general, and the density of livestock on what is available should be reduced" (adapted from *Beyond the Rangeland Conflict* by Dan Dagget).

Here's the type of solution that many second-generation ranchers might propose:

"We ranchers know better than anyone how to manage the rangelands to keep them healthy because we've been doing it for generations. The best solution is the simplest: leave us alone. Let those who know the land work the land. Let free-ranging cattle do what free-ranging buffalo have always done."

How do these "solutions" strike you? If you were an impassioned environmentalist, would the ranchers' solution fix the problem as you see it? If you were a rancher, would the environmentalists' solution fix the problem as you see it? Hardly!

The Problem with Environmental Problems

It wouldn't be all that hard to solve environmental problems if people weren't part of the equation. But people are part of the problem-solving equation. Always. Until that changes—which it never will—you need to find solutions that work for others as well as for you. This book will help you find those solutions.

Trying to solve tough environmental problems like the rangeland situation can seem overwhelming and hopeless, but it needn't be so. It's all in how you approach the problem and the people who care about the problem. That begins with understanding the "people factor"—what makes you and others tick, and how you, as a problem-solving group, tick. With that understanding, you can effectively apply lessons learned from disciplined problem solvers who have come before you. In this book, we'll present those lessons in terms of a problem-solving road map that shows you how to get where you want to go.

In chapters that follow, we'll talk lots about the people factor, the problem-solving road map, and how to orchestrate the two in order to solve tough environmental problems. But before moving on, here are a few cautionary notes.

The People Factor

Many well-intentioned people work hard to solve environmental problems. That's the good news. The bad news is that most of these well-intentioned people solve very few environmental problems. Often, in fact, their efforts make the situation worse rather than better. Here are some of the main reasons why:

- They do not understand how they and others think and approach problems.
- They spend too little time ferreting out the real problem, issues, and challenges. When participants fail to agree on the essence of the real problem, they wind up trying to solve different problems.
- They enter into discussions with preconceived solutions in mind. When this happens, success is measured by whether or not their preconceived solution is implemented, not by whether the true, underlying issue is addressed.

- They lack true open-mindedness. They are unable or unwilling to examine their own perspectives and arguments as critically as they examine those of others.
- They fixate on finding The Right Answer. They try to find one mega, do-everything solution to a problem rather than breaking the problem into specific challenges and solving the challenges one at a time.
- They fail to recognize that perspectives can be shaped by very different forces—reason, emotion, faith, culture, science—and that no single force trumps the others in value or importance.
- They do not follow a problem-solving approach (road map) that both insiders and outsiders can follow and understand. When others can't see how a touchy problem is being addressed, they invariably distrust the ultimate solution.

The Disciplined Thinking Factor

Just as inattention to the "people factor" ensures problem-solving failure, so it is with "disciplined thinking." The antidote to undisciplined thinking (what most of us seem to specialize in when trying to solve environmental problems) is a problem-solving road map.

To give you a taste of how disciplined thinking can be guided by a problem-solving road map, let's revisit the rangeland problem discussed earlier. To begin, a good road map would first flush into the open the full spectrum of issues and challenges underlying the rangeland problem. Only then could we know where we are and what we're dealing with.

Once we know where things stand, the next order of business would be specifying the desired outcome (e.g., clean water, a livable wage, a healthy rangeland) for each of the identified challenges. Some of these desired outcomes would seem reasonable and desirable to all parties. These are the ones that we'd tackle first.

To try to achieve a healthy rangeland as a desired outcome, for example, the road map would direct us to use some techniques for generating strategies to obtain this outcome. Some of the generated strategies might initially seem crazy (e.g., increasing grazing), others might not, but all possible strategies would be valued equally until the merits of each were evaluated carefully and thoroughly.

The road map would then help us figure out which strategies are most promising, and would direct us to give them a test run to see how they work. If a strategy lived up to expectations, we'd implement it right away; if a strategy needed some tinkering to make it work better, we'd do the tinkering and give it another test run.

If our implemented strategies didn't succeed at yielding the desired outcome (healthy rangeland), the road map would direct us to try other promising strategies. We'd continue this step-by-step process until the desired outcome was attained.

That's all there is to it (almost).

How It All Works in the Real World

Disciplined thinking, and understanding what makes individuals and groups tick, are interconnected processes that need to be managed concurrently. Fortunately, these two problem-solving elements build on one another. For example, in breaking a problem into definable challenges, the road map coincidentally shifts participants' attitudes away from knee-jerk reactions, and this leads to a better understanding of how "the other guy" sees it. This recognition that the other guy is not entirely evil shifts the dynamic away from strident, party-line dictums and directs attention instead to imaginative strategies that focus on specific desired outcomes upon which everyone has agreed.

The interconnectedness of disciplined thinking and the people factor is sometimes described as "finding common ground." Dan Dagget, an effective problem-solving environmentalist who has wrestled with the rangeland problem for years, describes it exactly this way: "I go about finding common ground . . . by having people of diverse, even opposite points of view identify the goals each of them wants to achieve on the land. Then I encourage those apparent adversaries to work together to reach whatever of those goals they have in common."[1]

Dagget, in *Beyond the Rangeland Conflict,* goes on to profile some of the strategies that have been test run on different ranches, once ranchers and environmentalists agreed on some common goals. In some cases, strategies that would have drawn immediate outrage and condemnation by one side

---

1. *Beyond the Rangeland Conflict,* p. 6. Compare this "common ground" approach to those offered at the beginning of the chapter. Which has the greatest likelihood of yielding outcomes that work for everyone?

or the other (e.g., increasing rather than reducing the intensity of grazing), achieved results that everyone favored. In fact, the outcomes based on common ground sometimes proved better *for both sides* than the knee-jerk, preconceived, party-line solutions stridently pushed by one side or the other.

If the doubting Thomas in you thinks this problem-solving approach sounds a little too rosy, that's a good sign: tough environmental problems can't be solved if you're unwilling to question (respectfully) what others present as "the way." But questioning "the way" cuts both ways: *your* "way" needs to be subjected to questioning and analysis as well. That's not such an easy thing to accept.

Have Dagget and others solved the rangeland problem? They certainly have made headway on at least one desired outcome, but they'd be the first to admit that the problem is far from being solved. Like all big environmental problems, solving the rangeland problem means solving pieces of the problem, one at a time, one place at a time. In the arid West, that means the rangeland problem will be solved one ranch at a time, one acre at a time. There is no other option, and that's how it goes with all environmental problems: there are no problem-solving short cuts, there are no quick fixes, and there are no silver bullets. The answers to pressing environmental problems are found locally, by people just like you. That's why it's so critical that you develop great problem-solving skills: environmental problems aren't going to be solved unless people like you lead the problem-solving crusade. You can't wait for the other guy to do it. Repeat (and practice) the following phrase often:

"If it is to be, it is up to me."

Chapter 2

How to Avoid Strangling Other Problem Solvers

Using a problem-solving road map to tackle tough environmental problems protects you from backtracking and reinventing the wheel. It also helps you explain to others how a problem is being tackled, what has been done to date, and what needs to happen next. This transparency of approach is especially critical when a problem-solving effort is interrupted in midstream, when a new participant joins the process, or when people affected by the outcome are distrustful of the decision-making process.

A Few Realities

Virtually all environmental problems are solved by groups rather than by individuals. As a result, to have any reasonable chance of solving an environmental problem that you care about, you need to understand and work with others who don't see the world as you do. That starts with recognizing and accepting from the start that people approach problems in different ways, and that no amount of hand-wringing, arguing, reasoning, or proselytizing is likely to change that. That's just the way it is.

Given our inherent differences—some of which are patently obvious, some of which are not—a good place to begin any problem-solving effort is to openly acknowledge, discuss, and even embrace differences within the group. This is especially important if problem-solving participants distrust one another or harbor ill feelings.

Differences in how people approach problems become readily apparent as soon as people start working to solve a problem. In fact, it often appears that interested parties have nothing in common except similarly strong-minded convictions about how the problem should be solved.

When dealing with environmental problems, finding a solution that works is nearly impossible unless interested parties (1) agree on the essence of the *real* problem and (2) acknowledge and respect one another's differences.

These two essentials are harder to orchestrate than you might think. So why bother even trying? Why not avoid the headaches of working with others? Why not do it alone? There are three main reasons: First, people need to feel that they are part of the process before they are willing to buy into "solutions" wholeheartedly. That means that, if you want your solutions to work, you need to include other interested parties in the process.

The second reason for weathering the difficulties of working in a group (or working with opponents rather than against them) is that a problem-solving team—if managed properly—is more effective at solving problems than a collection of individuals working in isolation.

The third reason is one we've already broached: all environmental problems are created by or complicated by what we've called the people factor. Doing it alone is therefore not even an option, so there's no point in pretending that it is.

Values

Working productively and cooperatively with others begins with understanding and respecting their values. This is especially important if there is an element of antipathy or distrust within the group.

Not all differences in style, approach, values or perspective are immediately obvious, so be careful about trusting first impressions too wholeheartedly. First impressions about how lighthearted and likable a person may be often prove accurate, but first impressions about how easy it will be to work with someone are often dead wrong. Work styles and personal approaches to problem solving are not immediately obvious or easily deduced. Sources of potential conflict brought on by different styles or approaches therefore remain hidden until people have worked together for a while, or until they have talked honestly and openly about their styles, approaches, and the buttons that set them off.

Usually—but not always—people work least contentiously with those who have styles, perspectives, values, and approaches similar to their own.

That is not surprising, of course, because it is easier to understand where similarly minded people are coming from and how and why things can go wrong.

However, working with people of like mind can be somewhat dangerous because a self-congratulatory comfort level is likely to settle into the clique. This can lead to false assumptions that one approach is right and another approach is wrong. This black-and-white mentality leads to trouble.

Educating the Uninformed

Oftentimes, key players in an environmental problem have different takes on a problem, with some people not seeming to "get it." When this occurs, it's important to find the reason why. It may be, of course, that those "not getting it" are truly unaware; it's also possible, however, that they *are* aware but just see the problem in a different light than you do. It's also possible that *you* are the one who's not getting it. Examine *yourself* before jumping to too many conclusions!

When advocating awareness-raising events such as demonstrations and protest marches, you aim to educate the unknowing about a problem.[1] But who really is the unknowing? Who really needs to be educated? Could it be *you?*

Before proselytizing under the guise of "education," be sure that you are willing to be educated as well.

So, be honest with yourself: are *you* actively and respectfully putting in a good faith effort to understand where *others* are coming from? Are *you* willing to have *your* mind opened to new perspectives? Are *you* willing to consider that *you* may be the one who isn't getting it?

It's not all that easy, of course, to consider perspectives that differ from one's own with an open mind, because we naturally distrust those who look at the world differently than we do. Developers and environmentalists, for

---

1. By calling attention to a cause, protesters and demonstrators seek to mobilize those who aren't aware of the problem. In so doing, they seek to build their support base for change. Counterdemonstrators try to nullify this effort. Both demonstrators and counterdemonstrators are targeting inactive bystanders.

example, tend to distrust and dislike one another by default; likewise for animal rights activists and hunters, property rights activists and government agencies. The more charitable among us may dismiss those with contrasting views as clueless or uninformed (rather than as evil), but that doesn't help us work with them effectively. Tolerance of what we judge to be misguided thinking falls far short of true understanding and respect. And remember:

Bending others' minds to see the world as you do is not education, it's indoctrination.

"Education" opens minds, "indoctrination" closes them. Thoughtful problem solving flourishes through the former; cults flourish through the latter.

If you're not willing to consider alternative points of view with a truly open mind, don't expect other people to consider yours.

Who Are They, Who Are You?

Most people consider themselves easygoing and laid back, and most people are—so long as things go their preferred way. The real test of an easygoing approach to life, however, is how you feel and behave when things don't go your way.

Antagonistic tendencies in a problem-solving group can be defused before they develop if you find out about one another's styles, perspectives, values, and approaches before a project begins. This "getting to know one another" works best if group members are first honest and upfront with themselves about their own personality traits and work-style preferences. The Meyers-Briggs test and the Glazer stress control test (see exercises at the end of this chapter) are both effective at revealing individual traits and tendencies.[2] After taking one of these tests and digesting the results, you and the others in your group should share your personal profiles openly

---

2. Test givers commonly typecast those who take this and similar tests, assigning summary scores or personality descriptors. There probably is merit in this, but not in a group dynamics context. The purpose of having team members take the test is to identify specific work-related traits of each member, not to assign labels to people. When working in a group, knowing specific traits of coworkers is more useful than generalized descriptions of personality type.

and honestly. This helps problem-solving participants understand one another better so that different approaches and work-style preferences can be understood and accommodated.

Few people are accustomed to being open about their own tendencies or to revealing the buttons that set them off. It therefore is necessary to create a nonthreatening, nonjudgmental environment where people can be open with themselves and others. An easy way to do this is for everyone in the group to read a shared, neutral source on managing group dynamics. This chapter can serve that purpose.

Be aware that you and others in your group will have a hard time really believing that personality traits are created equal—that they are neither good nor bad, just different. Cultural influences have indoctrinated us to believe that some traits are better than others—that arriving early is better than arriving late, for example. But one person's truth can be another person's falsehood. For example, it is likely that at least one person in a group will view lateness as a cardinal sin. But it's also likely that someone else in the group will view chronic punctuality as an uptight neurosis. Members need to be upfront with one another about their individual differences and peculiarities before they butt heads. Taking care of this business *before* starting a project averts unnecessary frustrations and ill feelings.

But telling others what buttons they shouldn't push isn't quite what we're talking about. Accompanying this information should be an attitude that:

"Buttons" are baggage that *you* carry with *you;* they are not wrongs inflicted on you by others.

If you have a hard time swallowing that, try explaining this: why aren't other people set off by the same buttons as you are?

How to Alert Others to Your Quirks and Buttons

A good way to set the proper tone for sharing things that bother you is to preface your admission with the following:

> "I wish it weren't so, but for some reason _____ really bothers me. I don't like feeling that way, and I'm trying to learn to be more relaxed about it, so please be patient with me."

This is a constructive approach because the speaker acknowledges that his/her button is personal baggage, not a universal wrong.

Now consider the following, more common approach:

> "I can't believe that there still are jerks out there who always _____. They drive me crazy. Don't do it if you want to stay on my good side!"

This approach is trouble and you should avoid it.[3] Remember that buttons are in the eye of the beholder, not in the action itself. The following, contrasting reactions may help you believe this:

> "It really makes me mad when people are late for meetings. What's wrong with them? How can they be so irresponsible?"

> "It really bugs me when people are neurotic about time. Don't they have anything better to do than sweat over a clock? Why can't they lighten up a little? Get a life!"

Here are more constructive ways to explain the same buttons:

> "I don't know why I'm like this, and I wish it weren't so, but I get really uptight about people being on time. I'm trying to learn to be more relaxed about punctuality, but it's hard for me."

> "I've come to realize that my casual approach to meeting times and deadlines works fine for me, but it doesn't always work well for people I'm working with. I'm going to make a real effort to be more punctual but please be patient with me as I fight my natural tendencies."

This second pair of admissions is obviously more useful in building good working relations in a group because the tone shifts from blaming others to acknowledging that some of the problem rests within you.

More Realities of Working with Others

Working with others can be frustratingly slow, and this wears on people who are goal-oriented and want to get things done. The pace may be so slow, in fact, that you may be tempted to take on the problem alone, leaving the rest of the team behind. If the problem is simple and straight-

---

3. If this is how you feel about your biggest buttons, you need an attitude adjustment!

forward, solo problem solving might make sense. But consider the consequences carefully:

It's very difficult to reconnect with others once you've alienated them.

To flog that essential point yet again, few (if any) environmental problems are ever solved by one person working in isolation. Likewise, few environmental problems are solved successfully by teams composed of similarly minded people.

Diversity of ideas and approaches on a problem-solving team is a strength, even if it doesn't always seem that way.

The key to working with others—without strangling them—is understanding one another's strengths and inclinations, and then planning tasks accordingly. Approach the group as a whole in a similar way: rather than dwelling on and bemoaning your group's shortcomings, focus instead on what your group does well. This shift in attitude, from focusing on what's wrong to what's right, sets a course for how the group can be most effective:

Identify what you do well together and do more of it. Doing more good is at least as effective as doing less bad.[4]

Three Ways People Attack Problems

As noted above, it is highly desirable to have members on your problem-solving team who approach problems in different ways, but only if members of the team consider different approaches to be strengths rather than liabilities. Interestingly, almost everyone favors one of three approaches to solving problems—"common sense," "formulaic," or "out-of-the-box thinking." You and others working on the problem need to know who approaches problems in which way or you'll drive one another crazy. Don't

---

4. When used by groups and organizations to bolster team dynamics and productivity, this technique of emphasizing the positives is called "appreciative inquiry." It is one application of "force field analysis" (see chapter 7).

wait until you're in the throes of heavy problem solving to figure out who subscribes to which approach!

The "Common Sense" Approach to Solving Problems

Many people rely heavily on common sense and practicality to solve problems. Their approach (also known as heuristics) places a high value on experience, convention, and common knowledge. Proponents of this approach to problem solving aren't quick to try new or different things: They know what works and they stick with solutions they have used in the past. Traditionalists and older people are generally associated with this approach because they often place a premium on real-world experience.

The "Formulaic" Approach to Solving Problems

Other people rely heavily on highly structured rules, formulas, and protocols (algorithms) to solve problems. Proponents of this approach have exacting, quantitative, structured minds; they distrust seat-of-the-pants approaches. Engineers and accountants often feel comfortable with this approach because they value highly ordered, unambiguous approaches that yield definitive answers where nothing is left to chance.

The "Out-of-the-Box" Approach to Solving Problems

Other people rely heavily on creative thinking to solve problems. Their approach (creative problem solving) leans toward imaginative musings and new assemblages of ideas. Proponents of this approach are always looking for a better way—the sky is the limit. They do not believe that the "way it always has been done" is necessarily the best way, and they do not believe that difficult problems can be reduced to black and white formulas. Designers and artists often subscribe to out-of-the-box thinking because they're daring and willing to try new things. It's important to understand and appreciate the merits of each approach if you're going to work with others. As the problem below illustrates, those subscribing to the "common sense," "formulaic," and "out-of-the-box" approaches generate very different strategies for how an environmental problem might be solved.

The problem: "Water quality in the Pleasant River seems to be getting worse each year, but the source of the pollution is unclear. Some think the East Branch of the river is the main source, others think it's the West

Branch. We can't afford to hire a consulting firm to figure out which of the two branches is worse, so we're looking for advice on how to figure this out."

A "Common Sense" Approach: "Let's do what they did in Lewis Creek. If it worked for them it should work for us. No need to reinvent the wheel!"

A "Formulaic" Approach: "It's clear that we need to collect water samples for analysis using the protocol sanctioned by the Environmental Protection Agency (EPA). We then need to analyze the samples in a certified lab and compare the results against standards set by the EPA. This will dispel any uncertainty about whether we have a water quality problem in either branch of the river."

An "Out-of-the-Box" Approach: "Wait a second, there are all kinds of things going on here that we haven't considered. How about checking with fishermen to see which branch of the river has more trout? Trout need clean water, so the branch with fewer trout is probably the more polluted branch. Or how about checking to see where people swim? People who are financially well off have lots of swimming options; people who are poor have fewer. Let's check streamside parking lots to see if a greater number of pricey cars are parked alongside one branch of the river than along the other. Or maybe we could . . ."

As the example above illustrates, the three types of problem-solving approach yield very different strategies. That's why it's good to work with people who look at the world differently than you do.[5]

But let's be realistic: getting people to appreciate diversity of problem-solving approaches is a challenge in itself, for those who embrace one approach tend to be aggressively critical of other approaches. Formulaic problem solvers view common sense strategies as too uncertain; they view out-of-the-box strategies as untested, unquantifiable, and risky. To common sense problem solvers, formulaic strategies seem impractical and academic; out-of-the-box strategies seem off-the-wall. To out-of-the-box problem solvers, common sense strategies might work, but there probably is a better way. As for formulaic strategies, they're too simplistic and one-dimensional. These strategists don't believe that the world works that way.

---

5. One way to ensure diversity of approaches and ideas is to invite public participation at particular junctures in the environmental problem-solving process.

And so, take a deep breath and *listen* to what others have to say. Ideas that go against your grain should be taken seriously not belittled, for no single problem-solving approach is best for all types of problems. You and others working on the problem need to understand from the outset how your problem-solving approaches differ so that you can exploit your differences and get the best out of everyone. Talking openly and honestly about your personal problem-solving tendencies *before* frustration and disrespect raise their dysfunctional heads is a good way to start.

Fear vs. Confidence

A dose of humility—an ability to acknowledge that you don't have everything figured out—is healthy and necessary if you're to work effectively with others. A healthy dose of self-confidence is also necessary, however.

Think positively and confidently about your problem-solving capabilities. Believe that you needn't rely on someone else's ideas or expertise. Believe that you enrich the problem-solving process when you look at things in different ways. Believe that there are good solutions out there to every problem you face. Believe that you will find them.

This upbeat attitude may seem like wishful thinking, but the importance of attitude in problem solving is real: Confidence breeds openness and creativity, both of which are needed to solve difficult problems. Few environmental problems can be solved effectively without them.

It's ironic that we value people who stand out in a crowd but we try so hard not to stand out in a crowd ourselves. Fear of being different or appearing stupid is counterproductive but it's not easy to overcome, because fear is reflexive, not actively controlled. Bravery is something you can control, however, and bravery counteracts fear. For example, you won't make much headway trying to talk yourself into not being afraid of heights, but you'll make lots of headway deciding to be brave. Decide to climb one foot higher on the ladder than you did last time. Similarly, rather than remaining frozen in fear about talking in meetings (and feeling terrible about your noninvolvement), decide to be brave and say hello to at least one person in your next meeting. After that small victory, set the bravery goal higher: talk with two people in your next meeting. Continue this incremental process of specific brave steps to rid yourself of counterproductive fear. Replacing fear with bravery should be a goal of every timid problem solver.

Another way to manage your fear and build confidence is to compare your inadequacies against those who seem to be more confident than you. Are you more critical of yourself than you are of them? Do you begrudge others for speaking their minds or thinking new thoughts? If you respect others for having the courage to challenge the status quo, why don't you show yourself the same respect?

Reject the bogeyman of problem-solving—FEAR—fear of trying something new, fear of making mistakes, fear of challenging authority, or group wisdom, or the status quo—fear of looking stupid.

How Stress Affects Problem Solving

Fear can bring problem-solving efforts to a standstill, but other types of stress can do this as well. Too much stress of any type, good or bad, can overwhelm you to the point that tying shoelaces becomes an insurmountable task.

You might assume that stress is intrinsically bad—that it adversely affects your ability to solve environmental problems. Surprisingly, that is not quite the case. It is true, of course, that you'll eventually reach a stress level that is debilitating—where your brain crashes and you can no longer think coherently (fig. 2.1). Situations with too little stress can be almost equally debilitating, however: Imagine, for example, a high paying, cushy, guaranteed-for-life job placing widgets on a conveyor belt. The stress level of this job would be very low, but that wouldn't make you feel energetic, productive, creative, or happy. In fact, just the opposite would occur: You'd become bored, complacent, and lethargic, and your productivity, creativity, and motivation would plummet. Life would drag on endlessly, dragging you along with it. You'd have a tough time solving even simple problems.

Contrast this stress-free scenario with one that requires some decision making each day. Responsible decision making would add stress to your life, but (so long as it's not too much stress) you'd feel more energetic, productive, creative, and happy. You'd be a much more effective problem solver.

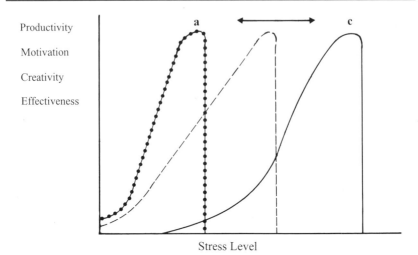

Fig. 2.1. How Stress Level Affects Productivity, Motivation, Creativity, and Enjoyment Person "a" works best in routine, low-stress environments; even moderate stress flips this person into a discombobulated panic. Person "c," in contrast, thrives in high-energy, stressful environments. This person is bored easily and is unproductive doing routine tasks where the level of stress is low. There can be too much of a good thing, however: if pushed too hard, even person "c" will crash into a worthless basket case.

Managing Stress So That It Works for You

You and members of your problem-solving team need to find and maintain your own personal stress levels at optimal levels, but you also need to recognize that coworkers probably have different optima and stress crash points than you do. Sleep deprivation, hunger, and other things in life—good stimuli and bad—push optimum stress levels and crash points one way or the other. You and other people working on a problem should be honest with one another about your stress comfort levels, and tasks should be planned accordingly.

One common source of stress on problem-solving teams is different attitudes about when things need to get done. Some people can always be counted on to complete tasks before they need to; others can be counted on to wait until the last possible minute before doing anything. Most of us have grown up with the "truth" that finishing early is good and that procrastinating is bad. But is that "truth" really true? Certainly some procrastinators are lazy and irresponsible, but what about those procrastinators who obviously do care about their performance, who work hard when they do work, but do not get around to doing anything until the last possible moment? Is their style worse than the early finisher? These procrastinators probably have high stress optima and they probably favor creative problem-solving approaches. Doing things too far in advance is boring and unproductive for them; they are not stimulated until the stress level is high. Tasks are subconsciously put off until their self-imposed stress level reaches a more optimal level.

For your problem-solving group to work well together, you must be honest and respectful of one another's stress optimum.

Be clear about your stress optimum before a project begins; this always works better than waiting until the pressure is turned up.

Let people with high stress optima handle last-minute tasks; let those with low stress optima (fig. 2.1) work on tasks that can be completed in the early stages of the problem-solving process. As long as it doesn't impede the work of others, procrastinators should not be pushed or expected to complete tasks ahead of deadlines; it is as pointless and unfair as exhorting early finishers to complete tasks by pulling all-nighters.

Staying out of the panic/basket-case zone is key, but minimizing stress is not the answer. People are happiest, most productive, most motivated, and most creative when they near their preferred stress level. But each person's stress optimum shifts, even daily, according to what else is going on in life. Very complex problems are best tackled by those who have high stress optima but do not have other stressful events consuming their energy. Others are likely to be overwhelmed unless assigned very defined tasks.

A Cautionary Note

This chapter has attempted to impress upon you that environmental problem solving is almost always a group endeavor and that "the people factor" must be taken seriously. Before moving on to the next chapter, one additional people factor needs to be acknowledged: motivation.

One of the most discouraging realizations that environmental problem solvers face every day is that you can't count on people to share your level of commitment or outrage. Many people just don't seem to care, or, if they do care, they don't translate their caring into action. That's plenty depressing, but there is a thin silver lining to this: at least these *non*doers aren't expending energy trying to *un*do your problem-solving efforts.

In later chapters, we'll suggest ways to make the best of the doer/nondoer/undoer reality by showing how the energies of doers can be focused, how the negative energies of undoers can be neutralized, and how the complacency of nondoers can be shaken. In the meantime, give some thought as to how the solving of your problem (page 1) would be affected by apathy (nondoers), focused positive energy (doers), and focused negative energy (undoers).

Exercises

Return to page 1 of this book where you recorded an environmental problem that you'd really like to see solved. Reread your description of the environmental problem. Now, if you can, find several people who also would really like to see this problem solved. Work with these people, as a group, on the "people factor" exercises below.

1. Who are you really? To reveal your personal work-style preferences, rank yourself along each continuum on the scale below. Ask coworkers to do the same. Talk openly with one another about your results.[6]

1	2	3	4	5

Always leaves things half finished	Always finishes what's started
Always early	Always late
Not competitive	Highly competitive
Never interrupts	Always interrupts
Always in a hurry	Never in a hurry
Takes one thing at a time	Does several things at a time
Expresses feelings openly	Holds feelings in

6. Adapted from "The Glazer Stress Control, Life Style, Self-Evaluation Questionnaire," Howard I. Glazer, Clinical Associate Professor at Cornell University Medical College/New York Presbyterian Hospital.

Focuses on practical	Focuses on the cerebral
Loves technology	Hates technology
Easily offended or upset	Not easily flustered
Defensive when challenged	Likes to be challenged
Never sets own deadlines	Always sets own deadlines
Detail oriented	Focuses on the big picture
Works best under pressure	Works best on own schedule
Always meets deadlines	Always misses deadlines
Likes to write	Hates to write
Hates public speaking	Likes talking to a crowd
Self-directed	Likes others to spell out tasks
Very concerned about what others think	Oblivious to what others think
Prefers to work with others	Prefers to work alone
Never judges in terms of numbers	Always judges in terms of numbers (how many, how much)
Needs to finish early	Waits until last minute
Needs to talk things over before acting	Anxious to get on with it
Always challenges authority and status quo	Never rocks the boat
Social interactions very important when working on project	Getting a good final product is all that matters
Agonizes over decisions (needs to consider all angles)	Makes decisions quickly, moves on, and doesn't look back

2. Summarize on paper your approach and work-style preferences—who are you? Be specific—address noteworthy elements from #1 above as well as other noteworthy traits that you can identify.

3. Compare and contrast your work-style preferences with those of people you live or work with. Which styles do you see as "good"? Which styles do you see as "bad"? Now play devil's advocate: how could traits that you judge to be good be seen as "not so good"? How could traits that you judge to be bad be seen as desirable?

4. Describe a person whose work style annoys you. Now describe three

traits this person possesses that might make him/her a valuable addition to a problem-solving team.

5. How does your own problem-solving approach tend to lean—common sense, formulaic, or out-of-the-box? Describe a difficult situation that you have tried to change (a problem that you have tried to solve) and how you went about it. How successful were you at solving the problem?

6. Think of someone who rubs you the wrong way. How does this person lean—common sense, formulaic, or out-of-the-box? Are this person's leanings different from your own?

7. Think of someone with whom you have strong disagreements about something important. Identify common ground that you share—basic elements about the issue or situation that you could agree upon.

8. Describe your "buttons." Why do you react to them as you do?

9. What buttons do you tend to push in those you live or work with? How would those people describe your button-pushing actions?

10. Describe situations that tend to lower your stress optimum.

11. Describe situations that tend to shift your stress optimum to a higher level.

12. Describe a situation, and the events leading up to it, where your stress level hit the panic zone. If you had it to do again, what would you do to stay out of the panic zone?

13. Describe a situation, and the events leading up to it, where a friend or family member hit the panic zone. What triggered the crash?

Applying Your Smarts to Real Problems

14. Each home on Mars Hill has its own septic system, but many of the systems are starting to fail. Suggest strategies that a homeowner might suggest or endorse if s/he:
   a. subscribes to the "common sense" approach to problem solving,
   b. subscribes to the "formulaic" approach to problem solving,
   c. subscribes to the "out-of-the-box" approach to problem solving.

15. The "Thrashers and Trashers," an all-terrain-vehicle (ATV) club, creates and maintains trails on state, federal, and private land for members' use. Other recreationists (hikers, mountain bikers, snowmobilers, and cross-country skiers) have been using the trails with higher frequency, and interactions with the ATV club members have become nasty, threatening, and sometimes physical. Your three-person problem-solving team has agreed to tackle this problem.

    a. What strategy might your "common sense" teammate suggest?

    b. What strategy might your "formulaic" teammate suggest?

    c. What strategy might your "out-of-the-box" teammate suggest?

Challenges for the Courageous (Don't attempt the following exercises if you've got everything figured out and are resistant to personal change!!)

16. So what's *your* problem? Consider each of the following traits and check those that describe how you think or behave.

Attitudinal Traits

- To your way of thinking, "open-mindedness" is the same thing as letting others have their ideas while you have yours.
- You're not the type of person who challenges authority or the status quo.
- When things don't go your way, you often say or think, "Just my luck!"
- You talk about others a lot, focusing mainly on things about them that you don't like or that you don't approve of.
- When things aren't going well, you instinctively look to blame someone else.
- You complain a fair amount.
- You try to act confident and together . . . (it's a good thing people don't know what's really going on inside you!)
- Careful and attentive listening are not your strong suits.

Interpersonal Traits

- You don't think very highly of people who push your buttons (you tend to think of them as jerks).

- You tend to hang around people who are negative and complain a lot.
- As far as you're concerned, proving someone wrong is almost as good as proving that you are right.
- To be honest, you don't always treat others the way you'd like them to treat you.
- Whether or not you have a good working or living relationship with someone depends on whether or not you like the person.
- You doubt that you ever could really respect someone who has political or ideological views entirely different from your own.
- You don't work as hard as you could at trying to remember people's names.
- Sometimes you don't get around to letting people know when you're going to miss or be late for an appointment.
- When working on a group project you oftentimes need to take over to get stuff done.
- In conversations, you tend to focus most on what *you* want to say (not so much on what others are saying).

Thinking and Problem-Solving Traits

- When trying to deal with several problems at the same time, you often get so overwhelmed that you can't do anything.
- You take it personally when someone challenges your ideas, comments, or beliefs.
- You don't have all that much confidence in your ability to figure out things on your own (without help).
- You tend to generate and evaluate ideas at the same time.
- You're not sure what the difference is between being a critic and being a critical thinker.
- When criticized, you tend to become mad or defensive (or you tune out what the moron is saying).
- You don't seem to make much headway on problems that really matter to you.
- When presented with a new, offbeat idea, your immediate reaction (whether you say it aloud or not) tends to be: "That won't work" or "That's not possible" or "That's a dumb idea!"

- You have a hard time believing that someone like you can really make a difference in the world.
- You tend to get angry or tune out when someone goes against something that you believe in or think you know.
- You're not very good at dissecting and finding flaws in other people's arguments (it's really frustrating to know that someone is talking garbage but not be able to counter the argument intelligently).

Take a moment now to reflect on each diagnostic trait that you checked. Place an "x" next to each that is obviously undesirable (a trait that you recognize as less than ideal); place an asterisk next to traits that strike you as inconsequential or innocuous.

What you now have before you is:
  a. a collection of ("x"ed) traits that you know ought to be changed,
  b. a collection of (asterisked) traits that you don't think are all that bad.

The traits that you recognize as undesirable (those in the first group) are relatively easy to work on and make better because you recognize the value of doing so. That's not the case with the traits that you deem innocuous: since you don't see them as undesirable, you're not motivated to change them (and you therefore won't change them) UNTIL something exposes them for what they really are. (If and when that happens, you may be surprised at how many traits you add to your "undesirable" list).

17. Review the collection of traits (#16 above) that you did not identify as problematic (the "b" collection). Be on the lookout for traits about which you react defensively—ones about which you find yourself thinking: "It's perfectly reasonable and understandable to act or think this way when someone does _____!" (or) "Are you saying I shouldn't stick up for what I think is right?!"

Visceral reactions such as these indicate that you are unable or unwilling to question or challenge your own beliefs, values, and "truths" as vigorously as you question and challenge those of others. Is this a trait that you admire in others? Is this a trait that you admire in yourself?

18. What change in each of your undesirable traits (the "a" collection from above) needs to occur for you to feel that it's no longer a problem?

(Record a specific, desired outcome for each trait that you wish to be different). For example, if you identified the first trait above—"To your way of thinking, 'open-mindedness' is letting others have their ideas while you have yours"—as an attitude that you'd like to change (a "problem"), you might identify your desired outcome as: "I truly am open-minded—I examine my own ideas and values as vigorously and critically as I examine those of others."

Chapter 3

How to Think More Like Albert Einstein and Less Like Homer Simpson

The first two chapters addressed "the people factor" and how to deal with it. Chapters 3 to 8 address the other half of environmental problem solving, "disciplined thinking." Disciplined thinking will help you think more like Albert Einstein and less like Homer Simpson.

Disciplined thinking is often used to mean both decision making and critical thinking. It has a broader meaning for environmental problem solvers, however, including the vitally important framework for attacking the problem. A framework (or "problem-solving road map," as we refer to it in this book) keeps you on course so that you don't waste time spinning your wheels, driving in circles, or trying to reinvent the wheel. The problem-solving road map also provides regular reminders to attend to the people factor.

But first things first—do you know a problem when you see one?

How to Know a Problem When You See One

Knowing a problem when you see one probably seems like the last thing you need help with. Homer, to the extent that he's able to think, would probably agree. Einstein wouldn't.

Situations that are unacceptably disturbing or upsetting certainly are *bona fide* problems, but so are daily chores such as starting the car, putting on socks, or turning on the shower. These everyday "things you do" aren't generally recognized as problems until something goes wrong—the car doesn't start, for example. You then need no convincing that you have a problem on your hands. The problem existed long before the car didn't start, however.

Relationship problems, financial problems, putting on socks, starting the car, and polluted water are problems because all are situations that you want to change and make different from what they are. In all cases you seek change in the status quo—from a fractious relationship to a harmonious relationship, from debt to savings, from bare feet to covered feet, from no shower water to flowing shower water, from a dead car to a started car, from dirty water to clean water.

Defining problems in this way—as situations that you'd like to be different—might seem a silly matter of semantics, but it's not. All problems—big, little, easy, difficult—can be approached the same way. If you understand how you solve simple problems, you can use the same approach to solve very difficult problems. Understanding that simple reality was part of Einstein's genius.

Try to think of "problem" this way:

A "problem" is any situation that's not quite the way you'd like it to be.

This all-encompassing definition includes "bad" problems, of course, but it also includes decisions, challenges, and welcome opportunities, such as having multiple job offers but not knowing which one to accept, or having too many volunteers step forward to clean up a river.[1]

Thinking of "problem" in this more expansive way is a requisite first step to solving environmental problems. Why? Because it focuses attention on what you want as a desired outcome—what a situation needs to look like for you to feel that the problem has been solved. After all, if you don't know what constitutes problem-solving "success," how can you know if you've achieved it?

Problems and the Beholder

Solving a tough problem ultimately depends on being able to define the problem clearly and accurately. Unfortunately,

1. In the case of "good" problems (not knowing which job offer to accept, for example) there's a disconnect between your current situation (indecision) and your desired situation (a good decision).

Few (if any) problems are absolute or interpretable in only one way: the true essence of a problem rests in the eyes of the beholder.

Consider, for example, two individuals who hold opposing views on abortion. Each has a strong but different sense about what's right and wrong, and each has a clear idea about what needs to happen to solve the abortion issue.

- The anti-abortion advocate may view abortion as murder of unborn children. To this person, the solution to the abortion problem is obvious: abortions (i.e., murder) must cease.
- The pro-abortion advocate may view abortion and accompanying legislation as an issue of women's rights. To this person, the solution is also obvious: abortions (i.e., what women do with their bodies) must be their choice alone.
- A mutual friend of the two may view the "abortion problem" even differently—as two good people not being able to get beyond their philosophical differences. To this person, the solution centers on human relationships rather than on abortion: having the two people get beyond the abortion issue so that they see the personal qualities in one another.

Environmental "problems" are also interpretable in very different ways:

Drilling for Oil in the Arctic National Wildlife Refuge (ANWR)
- Minimum-wage workers (and other consumers as well) may view drilling in ANWR as a way to lower gas prices at the pump. From this perspective, tapping into ANWR's oil reserves is desirable (a good solution to the problem of high gas prices).
- Environmentalists may view ANWR drilling as an assault on wilderness and the environment. From this perspective, drilling in ANWR is indefensible; the possibility of saving a few pennies at the gas pump does not justify the risk to ANWR's natural ecosystems.
- Alaskan politicians may view drilling in ANWR as a way to create jobs and generate tax revenue. From this perspective, the economic benefits might outweigh the environmental costs.

A National Organization Has Launched a Fund-Raising Campaign to Oust Politician X from Office.

- For politician X, this is a troubling development because he likes his job and doesn't want to lose it. From his perspective, the campaign is an undesirable situation (a problem), and its effects need to be nullified.
- For members of the organization, politician X's positions on the environment and business are undesirable (the problem). From their perspective, a "favorable" change in political positions is what they're ultimately after.
- For citizens who aren't attentive to politician X's positions, or who don't care one way or the other, or who hold all politicians in low regard, the campaign to oust him from office is of no concern (not a problem).

There Is a Proposal to Extend the Town's Water Line to Freemont Heights

- To Freemont Heights homeowners who have chronic water shortages, extending the town's water pipes is a solution to what they perceive as an insufficient water supply problem.
- To anti-sprawl advocates, extending the town's water to Freemont Heights is a problem, for it will lead to development of the surrounding agricultural fields.
- To residents from other parts of town who think that taxes are already too high, extending the water line is a problem because it will raise their taxes even further.
- To residents whose manicured lawns will be torn up when the new water lines are installed, extending the lines is a problem because it will ruin their yards.

As the examples above illustrate, people are very individualistic in how they see problems: one person's environmental problem is oftentimes another person's solution. So,

Never assume that your assessment of a situation (your take on the problem) is the same as everyone else's!

Different Ways to Solve Problems

There are many ways to go about solving problems or making decisions, of course, and no one way is universally good, bad, best, or worst. Some people trust intuition above all other approaches, others follow their emotions, others deal with problems and decisions by avoiding them, others rely on faith, and others rely on disciplined, rational analysis. Some of us rely on a combination of approaches, where the influence of each varies according to the type of problem or decision we're trying to address.

If a problem or decision affects you alone, you can proceed any way you wish, for you are beholden only to yourself. If the problem affects others, however—as do all environmental problems—you do not have that luxury: at minimum, you must be able to provide a clear, rational explanation of how and why you reached the decision that you did. Your explanation of your problem-solving or decision-making approach must be reasonable, easy to follow, and completely transparent. Self-indulgent explanations (e.g., "it's what I felt like doing" or "it seemed like a good idea" or "I had a hunch") may satisfy you, but they won't satisfy anyone else!

How Do You Approach Problems?

To reveal tendencies in how you deal with environmental problems and decisions, consider the following situations. Which approach—intuition, emotion, faith, avoidance, or disciplined, rational analysis—would you be most inclined to rely upon for each problem? (Be truthful!)

Situation 1a: There's a proposal within your town to build a new elementary school on the shores of Trout River (which is just upstream from where you live). Should you vote for or against the proposal? (You'd be most likely to rely on _____ to solve this problem and make a decision.)

Situation 2a: The pastor of your church or synagogue told parishioners how they should vote in the upcoming election, and the advice was far from centrist. Should you follow his/her advice? (You'd be most likely to rely on _____ to solve this problem and make a decision.)

Situation 3a: Wilson's Funeral Parlor wants to build a crematorium on the empty lot next to your home. How should you respond to this? (You'd be most likely to rely on _____ to solve this problem and make a decision.)

Situation 4a: The owner of the consulting firm you work for has asked you to go back to the site and redelineate the client's wetlands—"but this time, be less of a stickler or our client's permits won't be approved, and you may be out of a job." Should you listen to your conscience or should you listen to your boss? (You'd be most likely to rely on _____ to solve this problem and make a decision.)

Situation 5a: Six months ago you were in a meeting with this same bunch of people and some of them (on the other side of the issue) said things about you that were hurtful and untrue. Now the roles are reversed—you have the power to give them a taste of their own medicine. Should you do it? (You'd be most likely to rely on _____ to solve this problem and make a decision.)

How wise or appropriate were the problem-solving approaches that you relied upon? If you were to counsel someone who is confronting the same situation today, which problem-solving or decision-making approach would you recommend (i.e., intuition, emotion, faith, avoidance, or disciplined, rational analysis)?

Situation 1b: _____ probably would have been the best problem-solving or decision-making approach.

Situation 2b: _____ probably would have been the best problem-solving or decision-making approach.

Situation 3b: _____ probably would have been the best problem-solving or decision-making approach.

Situation 4b: _____ probably would have been the best problem-solving or decision-making approach.

Situation 5b: _____ probably would have been the best problem-solving or decision-making approach.

How Would You Like Others to Approach Problems?

The five situations above were similar in that you alone were the decision maker, and you had a higher stake in the decision than anyone else. The following situations are different. Your stake in the outcome is still high, but your role in the problem-solving and decision-making process is now low. Which approach—intuition, emotion, faith, avoidance, or disciplined, rational analysis—would you want decision makers to rely upon to represent your best interests? Which approach would you *not* want the decision makers to rely upon?

Situation 6: Your country's president is trying to decide whether or not to wage war on another country. (You'd hope that the President *would rely on* _____ to solve this problem and make a decision. You'd hope that the President *would NOT rely on* _____ to solve this problem and make a decision.)

Situation 7: The town's Conservation Commission will make a ruling on whether or not to approve placing the new landfill in your backyard. (You'd hope that the Commission *would rely on* _____ to solve this problem and make a decision. You'd hope that the Commission *would NOT rely on* _____ to solve this problem and make a decision.)

Situation 8: The power company wishes to install a high voltage power line through your neighborhood. Three of the five members of the commission that will vote on the power company's proposal used to work for the power company. (You'd hope that the commission *would rely on* _____ to solve this problem and make a decision. You'd hope that the commission *would NOT rely on* _____ to solve this problem and make a decision.)

Situation 9: The new park manager is deciding whether or not to uphold her predecessor's policy on trail use in the park. The current policy allows horseback riding but prohibits mountain bikes and ATVs (all terrain vehicles) from using park trails. Should the new manager change park policy? (You'd hope that the manager *would rely on* _____ to solve this problem and make a decision. You'd hope that the manager *would NOT rely on* _____ to solve this problem and make a decision.)

Situation 10: To cut costs and be more competitive, the Ticonderoga Mill is now burning old tires to generate power for the mill. You and other family members work at the mill, but you also live downwind of the mill. The Air Quality Division (AQD) of the State Agency of Natural Resources is reviewing the situation and will rule on it next month. (You'd hope that the AQD *would rely on* _____ to solve this problem and make a decision. You'd hope that the AQD *would* NOT *rely on* _____ to solve this problem and make a decision.)

Having considered the five situations above, were there any problem-solving/decision-making approaches that would serve your interests best? Were there any approaches that would seem arbitrary or unfair to you? Probably so! Unless you had unbridled confidence in the people making the decisions, you would be outraged if their decisions were based solely on emotion, faith, intuition, or avoidance. You would be much more willing to trust and accept their decisions if they were reached by disciplined, rational analysis.

This works both ways, of course: others will be most willing to trust and accept *your* decisions if you reach them through disciplined, rational analysis. So,

Opt for disciplined, rational analysis over other problem-solving approaches when the outcome of your problem-solving effort will affect others.

Remember that, just like you, people need to know *how* and *why* a chosen decision was reached, and they need to believe that the decision-making process was fair. Disciplined, rational analysis provides this, the other problem-solving approaches do not.

To summarize, if a problem-solving or decision-making effort affects you alone, then your choice of a problem-solving approach depends solely on what you have the most faith in; how you go about solving the problem or making the decision is nobody else's business. However, if the problem-solving or decision-making effort involves or affects other people (some of whom may have different views than you), disciplined, rational analysis is the only tenable way to proceed. For better or worse, that describes all environmental problems.

A Pep Talk from Your Coach

Unfortunately, learning to solve tough environmental problems isn't as easy as casually reading through a book. Like any other skill, appreciable improvement comes only through directed practice, the more the better.[2] There are no shortcuts.

Think of this book as your coach. The overall plan for success is explained first, with each incremental skill modeled for you. The book then coaches you through drills and practice exercises to hone essential skills. The problem-solving road map shows you how all the pieces fit together to get you where you want to go.

Don't be surprised if you start getting antsy about working through drills and practice exercises; it's perfectly understandable that you're eager to start tackling real problems that are more meaningful to you. That's as it should be. Be advised, however, that in real-world environmental problem solving, a poorly crafted "solution" can be much worse than no solution at all. Be prepared before stepping onto the problem-solving battlefield. Those who aren't pay the price.

How should you prepare yourself? Going through the motions half-heartedly won't do it, skipping practice won't do it, and having the coach practice for you won't do it either. The coach can only show you the way. Getting there is up to you.

Expect to experience some frustration and discomfort as you work to become a better problem solver. Change never comes without pain.

If you reach a point where working through exercises seems a waste of time, think again: there's a reason why famous musicians continue to take music lessons, and why English professors continue to take writing workshops, and why Olympians continue to receive coaching, and why master chess players consume every book on chess that they can find. The expression "no pain, no gain!" applies to environmental problem solving as much as to any other set of skills. You cannot become an appreciably better problem solver without disciplined practice, but disciplined practice requires an element of humility that you still have lots to learn. That was true for Albert Einstein and it's true for you and Homer Simpson too.

Hopefully, you now recognize that this book cannot tell you the "right answers" to your problems and decisions, and it cannot solve your problems for you. No book or coach can do that. This book will show you, however, *how to find* the best solutions and make the best decisions. The rest is up to you.

---

2. Consider people who are great at what they do—how have they succeeded where others haven't? Natural ability certainly can play a role, of course, but without drive and determination, genetic gifts are squandered.

One last note: *Be patient.* You can't run effectively before you've learned to walk, so gain experience solving local and regional environmental problems before trying to tackle national and global problems head-on. Megaproblems, unfortunately, won't go away while you're developing your problem-solving skills, they'll still be there when you're ready for them.

Exercises

Consider each of the following situations from a range of perspectives, identifying at least four different ways people might think about the same "problem." Be sure to identify a stakeholder group that would be likely to hold each view of the problem. For example, if the problem is: Wal-Mart might be opening a new store in St. Pierre, four different interpretations of this problem might be:

- (Small business owners): They might see this as a problem because it might rob them of some of their business.
- (Low-income residents of the town): They might see this as a good thing (not a problem at all) because the store has very low prices.
- (The highway department): They might see this as a problem because it will clog up existing roads.
- (The mayor and town council): They might see this as both a problem (the store will promote urban sprawl) and a solution (the store will provide employment for the unemployed).

1. The Shelburne Road section of Route 7 is being widened to a four-lane highway.

2. President Clinton's moratorium on road building in national forests has been overturned by President Bush.

3. Americans pay far less for a gallon of gasoline than do Europeans.

4. To generate electricity, wind turbines (with 20-foot propeller blades) are being installed where the wind tends to be strongest—on ridge tops and shorelines.

5. In the United States, it's illegal to sell or use DDT as an insecticide to control mosquitoes. U.S. manufacturers are still selling tons of DDT in less-developed countries, however.

6. The government has proposed a new law that requires all drinking water to be treated with fluoride.

7. Zebra mussel, a nonnative species, has invaded Lake Champlain.

8. Return to page 1 of this book and reread your description of the environmental problem that you would like to see solved. Now, as you did for the practice exercises above, consider this problem from a wide range of perspectives. How else might others think about this same problem? (Identify as many different perspectives as you can, and be sure to identify the stakeholder groups that would be likely to hold each view.)

Chapter 4

"DOC'S KEY": A Road Map to Problem-Solving Success

The first three chapters alerted you to several different realities of environmental problem solving:

- that solving environmental problems is, invariably, a group activity; trying to "go it alone" doesn't work;
- that people can observe or experience the exact same situation and interpret it in entirely different ways;
- that people use a variety of different approaches (e.g., emotion, intuition, faith, rational analysis) to tackle problems; and
- that environmental problems always affect more than a single individual; they therefore need to be solved in a way that seems fair and transparent to all. Only rational analysis and disciplined thinking meet those criteria.

Collectively, these realities might be described as "the people factor."

This fourth chapter addresses two additional realities:

- that you subconsciously follow a "road map" of problem-solving steps every time you tackle a simple problem;
- that, once you understand how your subconscious problem-solving road map works, you can use it to solve very difficult environmental problems.

But first things first: problems themselves.

Types of Problems

Problems come in many shapes and sizes, but all can be described as fitting into one of three types. Since the type of problem affects how you go about solving it, it's worth being able to recognize each problem for what it is.

A *right-wrong problem* is the most straightforward type of problem because there is at least one decidedly right or wrong solution. Examples include math problems, puzzles, broken equipment, and standardized test questions. As varied as these problems may be, all share the similar trait of having a correct solution. The problem is solved when the correct solution is found.

A *rule-laden problem* is one where the problem solver must dutifully follow a prescribed formula, recipe, protocol, or legislated sequence of steps to solve the problem. This type of problem might also be described as "problem solving by the numbers." Examples include legal problems, land-use permitting, chemical analyses, recipes, and income tax preparation. In solving a problem of this type, you may have a preferred outcome in mind, but the final solution and outcome is, technically, neither right nor wrong. The problem is solved by following the rules properly.

An *unstructured problem* is the most daunting type of problem because there is no right or wrong solution, and there is no preordained set of cookbook rules to follow to solve it. Examples of this type include relationship problems, political decisions, personnel issues, and any other problem that involves people having different values or opinions—that is, most environmental problems. Unstructured problems are the ones that keep you awake at night; they also are the ones that overwhelm environmental practitioners.

To summarize:

- **You are dealing with a right-wrong problem if your problem has decidedly right or wrong solutions.**

- **You are dealing with a rule-laden problem if laws, rules, or policies dictate that you must follow a specific protocol to address the problem.**

- **You are dealing with an unstructured problem if (1) no rules, laws, or policies dictate that you must follow a specific protocol to address your problem; and (2) there is no clear "right" solution to the problem. When tackling an unstructured problem, your goal is to find the "best" solutions rather than the "correct" solutions.**

Here are two examples of each type of problem:

- You paid for a $3.59 item with a five-dollar bill and were given back a dollar bill, a quarter, and a dime. Did you receive the right amount of change?

- In assessing a site to determine if it qualifies as a wetland, you try to identify all plants growing on the site. You're not sure if your identifications are correct.

These are *right-wrong* problems because the solution is black and white—either you identified the plants correctly (or received the right amount of change) or you didn't.

- You just received a speeding ticket for driving 40 mph in a 30 mph zone. Let's say that you don't want to pay the $100 fine, and that you don't want three points deducted from your license.

- You are evaluating a site to determine if it meets the "wetland" criteria as defined by federal statute.

These are *rule-laden* problems because there are prescribed guidelines set forth by authoritative bodies for how you go about determining if a site qualifies as a wetland (or how you go about challenging a traffic ticket). To solve either problem, you need to follow a prescribed protocol.

- Dan is always late for meetings and he often misses deadlines; it's really getting to the rest of us. We need Dan's contributions—he has talents that the rest of us lack—so we can't ignore him or write him off, but it doesn't seem fair or right that the rest of us meet obligations but Dan doesn't. We've tried talking with him about it, but he's emotional, easily upset, and extremely defensive. Unpleasant interactions make him sullen and confrontational, a mood that can last for days. We have a new project coming on board and Dan expects (and wants) to become the project leader. On paper, Dan would be a good choice, but in practice, we think he'd be a disaster. We don't know what to do.

- Loggers want to harvest trees from a virgin stand of timber; some environmental groups oppose any type of harvest.

These are *unstructured problems* because there's no single solution that all parties would agree is the right answer; there's also no prescribed formula for how to go about trying to solve the problem. Those who are trying to solve the problem are on their own.

Right-wrong and *rule-laden* problems may be challenging, but they seem easy when compared to *unstructured* problems. Unstructured problems lack rules, prescribed protocols, and clearly right or wrong solutions. There are no guidelines for what you should do, or how you should do it. Moreover, you may not even be sure when or if you've found a good solution.

Unstructured problems are messy and complex, with the complicating factor usually being *Homo sapiens*—you, me, and everyone else. The unpredictability of our individualities throws wrenches into otherwise straightforward problems. For example:

• A real estate developer approached John, our brother, about selling off the family farm. John is all for it and expects my sister and me to pay him cash for one third of the farm's projected worth if we don't wish to sell. The farm is mortgaged to the hilt, my sister and I don't have that kind of money, and we're not sure that our brother really is entitled to a one-third share anyway. The situation is becoming nasty and is tearing our family apart. What do we do?

• In the arid West, streamside vegetation zones protect water quality and spawning habitat for trout and provide habitat for rare species of plants. Streamside zones also provide abundant forage for livestock, however, and direct access to streams for thirsty livestock is necessary to make ranching viable. Strict rules governing the use of streamside zones are being vigorously advocated by some and are being vigorously opposed by others. As part of a team to solve this problem, what are you going to do?

As the examples above suggest, unstructured problems tend to be more troublesome and difficult to solve than structured problems. People often are stymied in their efforts to solve unstructured problems because they don't how to proceed. When they shoot from the hip, they sometimes make the situation worse.

The three types of problems—Right-wrong, rule-laden, and unstructured—are approached somewhat differently, so it's critically important that you're able to differentiate them. You need to know when you can, should, or must rely on a prescribed methodology to solve your problem. You also need to know when formulaic, cookbook approaches are inappropriate. You also need to be realistic about whether or not there is a right answer to your problem. In short, you need to be sure that your problem-solving approach fits your problem.

This book focuses on how to go about trying to solve unstructured environmental problems—the problems that are toughest to solve. As you will see, these types of problems can be made more manageable by imposing a little structure on them. The problem-solving road map described in this chapter will show you how to do that.

Dealing with Everyday Problems

You may not think of daily, familiar challenges as problems, but brushing your teeth, putting on socks, or tying a shoelace all represent situations that you'd like to make better (e.g., you want whiter teeth, warmer feet, shoes that don't fall off your feet). That is what makes each of them a bona fide problem. Solving these everyday problems is more involved than you'd think,[1] however, for there's a sequence of steps that you need to follow to get the outcome you want. For example:

When you wake up in the morning and feel that you want to make yourself more presentable before going out for the day, your awareness that something is not as good as you would like it to be activates your problem-solving thought processes and your quest to improve the situation. In short, you have identified a problem and want to do something about it.

Having recognized that you have a problem, you then somehow decide what exactly that problem is. All subsequent problem-solving steps and efforts are predicated on this first step of what you decide the problem to be.

Having zeroed in on a problem (something that you would like to be different), your mind naturally shifts to thinking about what you'd like as an end result. Your desired end result is your goal; when you reach the goal (get the result you want), you solve the problem.

Knowing what constitutes an acceptable outcome (i.e., your goal) is an important step in your subconscious problem-solving approach, but real action doesn't begin until you break your goal into specific, tangible objectives. When attempting to reach your goal of becoming presentable, for example, you naturally begin by making a mental list of discrete, desired outcomes. These desired outcomes (your objectives) represent targets you are shooting for—such as having clean, combed hair by 8 A.M., or having presentable teeth by 8:15 A.M.

---

1. Figuring out how to get socks on your feet may not seem like a problem because you know how to do it. Would getting socks on your feet seem more like a real "problem" if the socks were too small, or if you'd never seen socks before, or if you suddenly lost use of both arms and hands?

Objectives help you codify where you want to go, but they don't tell you how to get there. That's where strategies come in. Strategies are ways to get from point A to point B. When seeking ways to obtain cleaned, combed hair by 8 A.M., for example, you consciously or subconsciously generate strategies to meet this desired end. Since you have met this challenge many times in the past, you probably robotically implement tried and true strategies that have worked in the past.

When your problem strays from the familiar, finding a strategy that gives you the outcome you want isn't quite so simple. For example, if yesterday's hair dyeing adventure left your head looking ghastly, you may conclude that your default hair management strategies (e.g., shampooing, brushing) won't adequately solve your problem. You'll then need to search for other "answers" to your problem. In considering your different options, you'll unthinkingly place bounds on what strategies you are willing to consider. Strategies that will take too long, for example, or are too expensive, or are too risky, or are too socially unacceptable will not be considered.

Having placed bounds on the range of options you're willing to consider, you generate a number of potentially promising strategies, including "do nothing—leave the situation as it is." You then evaluate your options and select and implement whichever strategies make the most sense. After implementation, you evaluate how well your strategies worked. If you're not satisfied with the results, you backtrack and rethink the problem and possible ways to solve it. Alternatively, if you are wise and forward-thinking, you give your strategy a test drive before committing to full-scale implementation. The insights gained from this trial run suggest how you might modify the strategy to make it more effective. Oftentimes, a minor adjustment spells the difference between success and disaster.[2] A little troubleshooting can make all the difference in the world.

Once you're satisfied that your chosen strategy will give you the outcome you want, you implement the strategy. If for some unexpected reason your strategy falls short of the intended mark, you backtrack to find what went wrong and you try to fix it. If you can't, you head back to the drawing board and work through the problem-solving sequence again. You continue this process until you solve the problem to your satisfaction.

---

2. In some circles this is called "adaptive management."

The subconscious problem-solving approach described above exemplifies how any unstructured problem[3] can be solved. In the next section we'll use the same approach to tackle a more difficult environmental problem.

Dealing with More Difficult (Environmental) Problems

The subconscious problem-solving approach described above is how most of us subconsciously attack simple or familiar problems. This approach also can and should be used to attack complex, unstructured, or unfamiliar problems—that is, most environmental problems. You must be attentive to detail, however, to get the result you're seeking. The following example illustrates this.

> Let's say that your concern (problem) is the polluted state of a nearby river (see page xiii). You consider this a problem that needs attention because the condition of the river is different from what you would like it to be.

Having identified and defined this as a problem, you then consciously or subconsciously set a goal: to clean up the river so that it's not polluted. You then probably make some decisions about how the river needs to change to become satisfactorily "unpolluted." These desired end results (your objectives) might include outcomes such as reducing phosphorus levels, shutting down upstream factories, halting all direct sewage discharges into the river, and removing trash and junk cars from the river. With these objectives in mind, you might consider how your desired ends could be sabotaged by obstacles and constraints such as citizen apathy, insufficient funds, or town politics. The obstacles might cause you to give up the fight because you conclude that the deck is too stacked against you—you just can't win. Alternatively, if you are sufficiently outraged by the river's condition, you might attack the problem and obstacles with a vengeance. To do so, you seek out strategies to move the pollution problem from where it is now to where you want it to be in the end. You try out what you think are the best ideas and hope that they work. If they do, you

---

3. Note that the problem "I wish to make myself more presentable for the day" is an unstructured problem rather than a rule-laden or right-wrong problem. That's because there are no clearly right or wrong solutions to this problem and there are no legislated rules or prescribed protocols as to how you must go about solving this problem.

conclude that you have solved the problem. If your implemented strategies don't achieve the desired ends, you regroup, give up, or try another strategy.

The approach above has the following features:

• it seems reasonable and familiar to most people,
• it resembles how most of us would approach the problem,
• it probably won't give you the results you want.

Why? What's the matter? The failing is not the approach, it's in the casual execution of the approach.

Before reading further, try to identify weaknesses in how the approach was executed. Place your answers on the lines below. **Hint:** Read the above paragraph a second and third time, but this time read *carefully and thoughtfully.*

1. _____

_____

_____

2. _____

_____

_____

3. _____

_____

_____

How many weaknesses are there? Many. Here are two that make failure a near certainty:

1. Failure to be clear about the real problem. The *real* problem was not clearly articulated or defined. Was human health the real concern? Decline in trout populations? Smell? Appearance? These concerns (problems) all might be described as "pollution" problems, but solving these very different problems requires very different strategies. For example, if disease transmission from human waste is the real concern (problem), then shutting down factories or removing junk cars will do nothing to relieve the problem. Alternatively, if aesthetics is the real issue, then reducing levels of an unseen chemical—even if the chemical is toxic—will be ineffective. If the real problem is fewer trout, little will be gained by removing trash from the river.

2. Failure to be clear about what constitutes problem-solving success. Another major failing in the problem-solving effort described above is failure to identify what outcomes are needed to have the problem solved. The stated objectives, while sounding good, are only knee-jerk reactions to an unspecified problem; that's why the strategies (solutions) are bound to fall short. For example, the objective "to reduce phosphorus" is virtually meaningless because it fails to identify a benchmark for what constitutes an acceptable reduction in phosphorus. Would a one-molecule-reduction in phosphorus one hundred years from now be an acceptable outcome? Probably not, but without specific, clearly stated, measurable targets there is no way for problem solvers to determine if an implemented strategy (a solution) does its job.

To recap, efforts to solve difficult (unstructured) environmental problems are certain to fall short if you lack, or fail to follow carefully, a methodical plan of attack. Fortunately, you have such a plan of attack at your disposal—it's your subconscious problem-solving road map.[4]

If you're attentive to detail, you can use your natural, internal problem-solving road map to impose structure on tough environmental problems. Doing so ensures that your problem-solving thought processes are orderly and complete.

The Ins and Outs of Your Internal Problem-Solving Road Map

There are seven sequential steps in the problem-solving road map that you and other great thinkers naturally use to solve problems.[5] Conveniently, the pathway can be remembered as "DOC'S KEY" to problem-solving success (see also pages xiv–xv):

Definition: Clearly define the problem and your desired outcome (and distrust the obvious!).

Objectives: Set clear objectives that are specific, measurable, attainable, reasonable, and timely.

---

4. Remarkably, this road map varies little from person to person.
5. How-to techniques for each step in the road map are presented in chapters that follow.

Constraints: Identify constraints, boundaries, assumptions, limitations, and unacceptable impacts.

Strategies: Seek creative strategies—generate ideas.

Keepers: Select your best strategies.

Experiment: Try out your strategies (and make adjustments).

Yes! Go ahead and implement your strategies (and check to make sure they give you the results you want).

Here's a more detailed explanation of each step:

Step 1. *Definition*: Clearly Define the Problem and Your Desired Outcome

It may seem obvious that you need to figure out the true essence of the environmental problem you are trying to solve before you are likely to solve it, but this first critical step in problem solving is rarely given the attention it deserves. Instead, people tend to forego problem definition so that they can get on with finding solutions. This works fine, but only if you're not fussy about which problem your generated solutions might solve!

Accurately defining the real problem—spelling out what constitutes problem-solving "success"—is surprisingly difficult and time-consuming because the *real* problem is rarely what you initially think it is. In fact, if a problem seems obvious to you, you're probably on the wrong track. Ferreting out the *real* problem rarely comes that easily.

Defining a problem accurately is by far the most important step in problem solving for it guides the rest of the problem-solving effort. When you fail to identify the problem and desired outcome accurately, your "solutions" are destined to miss their desired target, and they'll fail to bring about the outcome that you really desire. When you solve the wrong problem, you're right back where you started—if you're lucky.

One common obstacle to problem-solving success is failing to distinguish preconceived strategies from bona fide problems. In fact, this is the single biggest reason that so many environmental problems become adversarial and unsolvable: when players become wedded to certain strategies without determining whether they address real "problems," there can be no common ground among the interested parties. The dynamic immediately degenerates into, "How do I get *my* strategy implemented?" rather than "What is the real problem?" and "What outcome is needed for us to feel that the problem has been taken care of?" and "What strategies might we employ to achieve that outcome?"

Here are some examples of preconceived strategies that are mistakenly treated as problems:

• Cattle grazing along streams must stop.
• We must enact legislation for land-use planning.
• Leghold traps should be outlawed.

Start worrying when you hear yourself or others use the word "must" or "should." These exhortations usually signal that a strategy has been pushed, not that a problem has been defined. Well-defined problems use neutral phrasing that describes the situation rather than a pet strategy. For example:

• When salmon eggs in Spring Creek riffles are covered with sediment they don't hatch.[6]
• Our current approach to land-use planning is not controlling urban sprawl.
• Animals caught in leghold traps seem to endure pain and suffering.

Step 2. *Objectives: Set Clear Objectives*

Clear objectives are essential in environmental problem solving because difficult problems are solved by meeting objectives, not by finding mega-solutions to the overarching problem.

Objectives must be stated explicitly so that you and others always know where you are going and how close you are to getting there. Useful objectives are specific, measurable, attainable, and reasonable; they also include an explicit timetable for completion.

Good objectives are obvious extensions of the defined problem and desired outcome.

The process of setting objectives is an essential part of environmental problem solving because it forces you to break complex problems into manageable challenges. This "divide and conquer" approach simplifies the problem-solving effort by providing clear targets for where you want to go. Setting objectives simplifies the problem-solving challenge to: "How might we get where we want to go?" or "What do we need to do to hit the targets?"

6. Note that these problem statements are much more open-ended than the preconceived strategies presented earlier, and that a wide range of possible strategies is now suggested.

Step 3. *Constraints:* Identify Constraints, Boundaries, Assumptions, Limitations, and Unacceptable Impacts

This step in environmental problem solving is less obvious than problem definition and objective setting, but almost as important. Think of constraints as reality checks—who is part of the problem and who is not? What are the geographical boundaries of the problem? Are there any time, money, or legal constraints that could limit what you might do? What impacts on the human or natural environment (resulting from implementation of a strategy) are not acceptable? Be explicit and discerning about problem-solving boundaries that might limit your pursuit of strategies. Also be vigilant for assumptions that may not necessarily be true. Faulty assumptions unnecessarily limit your problem-solving options.

Thinking about constraints and unacceptable impacts should shed new light on the real nature of your problem and objectives. Use this as a checkpoint in the problem-solving road map to adjust your problem definition (step 1), objectives (step 2), and constraints (step 3) so that they mutually reinforce one another.

Step 4. *Strategies:* Seek Creative Strategies—Generate Ideas

The first three steps of the problem-solving road map clarify what the problem is really about and what outcomes constitute problem-solving success. Now is the time to think creatively and to explore numerous ways to meet your desired ends—the more possible strategies the better! This works best when you refrain from judging the goodness or badness of ideas as you generate them. Evaluation comes later.

Step 5. *Keepers:* Select Your Best Strategies

After generating a bunch of possible strategies that might get you where you want to go, you (at some point) must decide which possible strategies are keepers and which are not. This is that time.

Step 6. *Experiment:* Try Out Your Strategies (and make adjustments)

Before unleashing your "keepers" on the world, are there confounding factors that could derail their successful implementation or effectiveness?

Can you forecast what possibly could go wrong? Can you do anything to minimize the damage? What are the worst-case scenarios? What unintended economic, cultural, social, human, and environmental impacts might result if you put your chosen strategies into action? What can you do to minimize their impacts? Troubleshooting is an effective way to make good ideas better—and avoid preventable crises.

Troubleshooting your keepers before implementation will reveal weaknesses and shortcomings. Whenever possible, however, also give your strategies a test drive before implementing them fully. Trying out your ideas in a limited way before full-scale implementation helps you identify weaknesses so that you can take corrective action while there still is time.[7] Very often the difference between a strategy succeeding or failing is in the details.

Step 7. *Yes!* Go Ahead and Implement Your Strategies (and check to make sure they give you the results you want)

If your implemented strategies give you the outcome you're looking for, congratulations! You've gotten where you want to go. If they don't give you the results you hoped for, however, it's time to regroup and figure out why. Begin by reconsidering how you defined the problem and the desired outcome. Are you *sure* that you ferreted out the *real* issue that bothered you? If so, did your stated objectives fully capture the outcomes you desired? What was missing? And how about constraints—did you make unnecessary assumptions that limited your range of options? As for strategies, did you come up with as many as you could (including offbeat ones), or did you settle for a few old standbys? And when you decided which strategies to keep, were your criteria for selection appropriate? Did you use your objectives as benchmarks when you evaluated the effectiveness of each possible strategy?

The seven-step road map above—"DOC's KEY" (**D**efinition, **O**bjectives, **C**onstraints, **S**trategies, **K**eepers, **E**xperiment, **Y**es!)—provides a problem-solving road map that works well for all "unstructured" and "right-wrong" environmental problems.[8] If the road map strikes you as contrived, try to

---

7. You wouldn't market a newly invented car without first trying it out to find the bugs, would you?

8. "Rule-laden problems," by definition, demand that you follow a cookbook problem-solving methodology. Such methodologies, while perhaps effective for the problems they were designed to address, rarely work well when applied to unstructured problems outside their domain.

withhold judgment until you have given it a fair try. With use, you will recognize that the road map is nothing more or less than a formalization of your subconscious problem-solving approach, with troubleshooting (experimentation) thrown in to keep you out of trouble. The road map works well because it dissects complex environmental problems into manageable pieces, while still providing a coherent structure that connects the individual pieces to the whole.

When you follow a problem-solving road map carefully, you are protected from backtracking or reinventing the wheel. You're also better able to explain to others how an environmental problem is being tackled, what has been done to date, and what needs to happen next. This transparency of approach is especially critical when a problem-solving effort is interrupted in midstream, or when a new participant joins the process, or when people affected by the outcome are distrustful of the decision-making process. Those complications tend to be the rule rather than the exception with environmental problems.

Exercises

Each of the following situations (1–14) could be considered a "problem." Identify several different but realistic perspectives on each problem, and who might hold that perspective. For example: Our city is considering an ordinance that would ban smoking in bars. Different perspectives on this might be as follows:

- To patrons who smoke: The ban would be a problem because smokers would not be able to do what they like to do.
- To bar owners: A ban on smoking would be a problem if it adversely affected business.
- To bar patrons who don't smoke: The ban would improve a situation that they have not liked; it would solve a problem that they've been tolerating.
- To the mayor: If smokers are sufficiently outraged, this could become a problem for the mayor, for s/he might be voted out of office; etc.

1. A rare species has been found to grow on a proposed development site.

2. Our town water is treated with fluoride.

3. A retired farmer in our town wishes to subdivide and develop his land but town laws and regulations say he can't.

4. An energy company wants to create a hydroelectric plant in northern Canada. To do so, it will be necessary to build a dam that floods thousands of acres of forest.

5. Many people living in Las Vegas and other desert locations demand lush, green lawns.

6. There's a proposal to prohibit dogs from the town parks.

7. The introduction of brown trout, a nonnative species, to McCabe Creek has helped the region's economy but hurt the long-term viability of the native trout population.

8. Environmental laws and regulations are less strict in third-world countries than they are in the United States, Canada, or Great Britain.

9. A moratorium has been placed on all road building on U.S. Forest Service land.

10. NEPA (the National Environmental Policy Act) mandates that any proposed activity on federal land that is likely to have a meaningful impact on the environment must follow the environmental impact assessment process.

11. Construction of the "billion dollar bridge" has come to a screeching halt because a rare turtle has laid eggs near the construction site.

12. American chemical factories are selling DDT and other banned insecticides to countries where environmental laws are less strict.

13. A road has been built through Victory Marsh to improve access to Mineral Springs, but the road has altered the hydrologic flow; also, permitting regulations were ignored.

14. Mom and dad, with no special skills or job prospects, both lost their jobs when EPA closed down the factory for pollution violations.

15. Return to page 1 of this book and consider the environmental problem that you described. Why hasn't this problem already been solved? Why hasn't solving it been as easy as getting socks on your feet? Where have past

problem-solving efforts gone astray? (Record your thoughts on the lines below):

_____

_____

_____

_____

_____

Chapter 5

Soapboxes, Bandwagons, and Real Problems: The "D" in DOC'S KEY

This chapter is about defining problems (the "D" in DOC's KEY) and knowing problems and solutions when you see them. That probably seems like the last thing you need help with, for you surely know what's a problem and what's not!—at least you think you do. Therein lies the biggest obstacle to effective problem solving:

Problems are rarely as they seem. Until you are able to imagine that maybe—just maybe—the _real_ problem is not quite what you assume it to be, you will spin and reinvent lots of wheels. You also will create as many problems as you solve.

To make this more believable to you, let's see how you would deal with the following challenges:

Challenge 1. You and some friends are upset about the cigarette butts that litter the town park. (There are ashtrays in the park, but some smokers aren't using them). It's obvious that smokers just don't care. You've decided to do something about this situation. What is that "something"? Think it over and write down your idea of a good solution on the lines below:

_____

_____

_____

Good! One problem down, only a few gazillions to go! Here's another warm-up before we get to the heavy lifting:

Challenge 2. I need a new car but I'm low on funds and I don't want to take out another loan—got any bright ideas? Think it over and record your wise thoughts on the lines below:

_____

_____

_____

Since you're on a roll, here's another real-world problem—help!

Challenge 3. Fur traders head to the Arctic each year to club baby seals to death for their fur. What can we do to stop this gruesome and cruel practice?

_____

_____

_____

How Wise Were Your Strategies?

Being able to come up with strategies for tough problems, without lots of dithering, is a wonderful skill—but only if the strategies work. Let's do a quick check on just how good your proposed strategies were. We'll begin by comparing your strategies to those offered by several hundred college students and townspeople who were presented with the same three challenges.

Here are the ten most common strategies for the cigarette butt problem (Challenge 5.1):

- *Strategy 1:* Put up signs where people congregate to smoke that explain the environmental pollution caused by cigarette butts.
- *Strategy 2:* Triple the number of ashtrays in the park.
- *Strategy 3:* Post signs all over the park to raise awareness about cigarette butts as litter.
- *Strategy 4:* Require all smokers to do community service.

- *Strategy 5:* Require all park users to pass a test demonstrating that they understand the effects that cigarette butts have on the environment
- *Strategy 6:* Distribute to all new town residents a brochure that addresses environmental ethics and courtesies
- *Strategy 7:* Create a "butt police" to patrol smoking areas and levy hefty fines on those who dispose of butts improperly.
- *Strategy 8:* Move existing ashtrays to where the butt problem is worst.
- *Strategy 9:* Force smokers to pick up their butts.
- *Strategy 10:* Cordon off designated smoking areas so butt litter is restricted to certain places where it's easily cleaned up.

Strategies along the lines of those above account for 95 percent of the strategies offered by those surveyed. If your proposed strategy resembles one of the above, you're thinking along the lines of the majority. Unfortunately, that's not something to be proud of. Solutions such as those above are actually very weak and ineffective. It's likely, in fact, that no combination of these strategies would give you the outcome you're looking for.

Most of us discover weaknesses in our strategies the painful way—we implement our chosen solutions and then watch them fall apart. A wiser, more effective approach is to think through the likely outcomes of strategies *before* implementing them. You then can decide—before committing to a course of action—which strategies are most likely to give you the result you want.

To illustrate how you might evaluate a range of possible strategies, let's forecast possible outcomes for each of the proposed strategies to the cigarette butt problem. We'll start with strategy 1—*Put up signs where people congregate to smoke that explain the environmental pollution caused by cigarette butts.* OK, let's imagine that you've put up a bunch of signs. What might be the outcomes of your effort? There are several:

a. Smokers read the signs and stop throwing butts on the ground.
b. Smokers read the signs and throw fewer butts on the ground.
c. Smokers read the signs but still throw the same number of butts on the ground.
d. Smokers read the signs but now throw more butts on the ground (maybe they're piqued by the signs and are showing it by increased littering).
e. Smokers don't read the signs (and their butt throwing behavior remains unchanged).

f. Passersby don't like having more signs in the park.

g. Passersby like having more signs in the park.

h. Passersby are neutral about having more signs in the park.

i. Existing butts on the ground are picked up.

j. Existing butts on the ground are not picked up.

Here's the forecasting test: How satisfactory (to you) are these possible outcomes?

- If outcome (a) is the one and only outcome, then you probably would feel that strategy 1 worked—that the cigarette butt problem was taken care of and it's no longer an issue.

- How about outcome (b)? If virtually all butt throwing ceased, you might feel that's good enough for you—that the problem (as far as you're concerned) was adequately solved. If only a couple of smokers stopped throwing their butts on the ground, however, and the remainder continued their undesired behavior, would you feel satisfied that the butt problem was solved? Probably not.

- How about outcomes (c) and (e)? Both would be disappointing and unsatisfactory.

- Outcomes (d) and (f)? Changes for the worse—hardly what you want!

- Outcomes (g) and (h)? Not bad outcomes in themselves, but if they're the only outcomes, the cigarette butt problem would continue to bother you.

- Outcome (i)? No other outcome would be satisfactory if this didn't transpire.

- Outcome (j)? Forget about it. If this happened, every other outcome would be a loser.

Having forecasted possible outcomes, what do you now think of strategy 1? Is it a winner or a loser? By forecasting the possible outcomes, you were able to take a first cut at assessing the likelihood that putting up a bunch of signs would take care of the cigarette butt problem. Given the large number of reasonable outcomes that would fall short of resolving the problem, it seems that putting up signs might not give you the results you want.

Let's now forecast possible outcomes for strategy 2—*Triple the number of ashtrays in the park*—to see if it's likely to yield the outcome you want. Putting out more ashtrays seems like a no-brainer—it certainly couldn't hurt and it probably would help. Is that your thinking? If so, you need to think more and shoot from the hip less. "Putting out more ashtrays couldn't hurt, and probably would help" is an ineffective way to try to change the cigarette butt situation. Why waste time and money and energy on a wishy-washy solution when you instead could implement one that's much more likely to give you the outcome you want?

Put on the brakes whenever you find yourself thinking, "We may as well implement 'solution x' because it can't hurt, and it just might help." This line of thinking indicates mental laziness. It also indicates that you haven't really figured out the *real* problem or desired outcome.

Here's what might happen if the number of ashtrays were tripled:
   a. Smokers would deposit all of their butts in ashtrays.
   b. Smokers would deposit some of their butts in ashtrays but throw others on the ground.
   c. Smokers would pay no attention to the new ashtrays (their tossing of butts wouldn't change).
   d. Passersby would not like seeing the park littered with more ashtrays (they might find the ashtrays more objectionable than the cigarette butts).
   e. Someone would vandalize the new ashtrays.

As you can see from these forecasted outcomes (and there are other possible outcomes as well, of course), tripling the number of ashtrays would not ensure that the cigarette butt problem would be resolved to your liking. That does not mean that you necessarily should abandon the ashtray idea as worthless, of course. Forecasting possible outcomes and identifying the ones that are not to your liking lets you anticipate—before it's too late—what might go wrong. This troubleshooting insight gives you a chance to tweak your strategy to avoid undesirable outcomes.

If you read and processed the preceding thoughtfully, you now should be more cautious about embracing "solutions" without first being clear about exactly what's bothering you, and exactly how the situation would need to be different for you to feel that it's no longer a problem.

To summarize, here's where we've been and where we are going: Exercises 5.1–5.3 introduced a more systematic way of approaching situations that are not to your liking. Working through these exercises demonstrated that "safe" ideas are not necessarily worthy ideas, and that problems and their "solutions" are rarely as straightforward or simple as they first appear. The real essence of effective problem solving hinges on clearly and accurately diagnosing the real problem before worrying about possible solutions. Einstein figured this out a long time ago:

If he had but one hour to save the world, Einstein said he'd spend fifty-five minutes figuring out what the real problem was, and five minutes finding a solution to it.

Unless you're smarter than Dr. Einstein, you should follow his lead.

From this point onward, we'll mostly avoid using "solution" and will use "strategy" instead. Here's why: "Strategy" and "solution" both refer to an *idea* that is implemented to solve a problem. "Solution," however, is sometimes also used to describe the outcome of a strategy being implemented. The resulting uncertainty of meaning can create confusion when following the problem-solving road map.

Some Unorthodox Cures for Cigarette Butt Litter

After forecasting possible outcomes for a couple of the proposed solutions to the cigarette butt problem, we then assessed how satisfied we'd be with each possible outcome. We discovered that the proposed solutions were not as satisfactory as they first appeared.

Forecasting outcomes and evaluating the acceptability of those outcomes helps you tackle problems more effectively. To illustrate, consider the following unusual strategies that might be employed to change the cigarette butt situation so that it's more to your liking:

• Regularly cover butts on the ground with mulch or flowers so that they're not visible.

• Convince the anti-butt vigilantes not to be bothered by the appearance of cigarette butts on the ground.

What do you think of those strategies?

Your first reaction may be incredulity that we're caving in to smokers who are slobs. Why should the rest of us be the ones who make changes when they are the ones causing the problem? Your second reaction is hopefully more appreciative . . . but that depends on whether or not you have identified the real problem and desired outcome.

In truth, your displeasure with the situation probably revolves around aesthetics, resenting smokers, or both. Perhaps the sight of discarded butts compromises your enjoyment of the park's beauty. Perhaps you view the discarded butts as statements that smokers care only about themselves. Perhaps the sight of discarded butts triggers memories of past run-ins with self-righteous smokers. If those categorizations capture the essence of the problem as you see it, then your desired outcomes are quite simple:
• You don't want to see discarded butts in the park.
• You want smokers to be accountable for their actions.

When you look at the situation in this way—focusing first on the essence of the problem and what constitutes an acceptable outcome—all kinds of unusual, imaginative strategies emerge—from giving psychotherapy to those who don't like cigarette butts (so that they learn to love the appearance of tossed butts), to cordoning off butt-littered places so that offended souls don't see the butts, to regularly covering up tossed butts with gravel, to showcasing collections of tossed butts as works of art.

Offbeat ideas? Maybe. Stupid ideas? Absolutely not. The proof is always in the pudding: If a proposed strategy gives you the outcome you want, the strategy is a winner, even if the strategy is unorthodox. Conversely, if the strategy doesn't give you what you want, the strategy is a loser, even if it is considered reasonable by everyone else. It's as simple as that.

Before moving on, we need to emphasize yet again that there's a giant difference between what you want in the end (your desired outcome) and how you'll make it become a reality (your strategy/solution). Your desired outcome is a reflection of the problem as you see it—it's the thing that needs to materialize for you to feel that the problem has been solved. Strategies/solutions are the means to that end. The solutions that you advocated for challenges 5.1–5.3 were "strategies"—propositions for what might be done to make a situation more to your liking. But were those strategies designed to yield specific outcomes, or were they quick reactions to what seemed a right thing to do? Did you figure out the essence of the real problem before seeking solutions? Were you clear about exactly where you wanted to go before you started trying to get there? To ask it another way, did you follow Einstein's lead?

Real Problems vs. Preconceived Solutions

You can avoid many headaches and frustrations by recognizing from the get-go that many of your proclaimed "problems" are really not problems at all. Here are some examples:

- "I need a new car."
- "We need to stop the cutting of tropical rainforest."
- "Clubbing baby seals must be stopped."

All of these are really preconceived solutions to hidden problems. The real issue has not been identified.

When you say "I need a new car," you're advancing a prefabricated strategy (getting a new car) rather than articulating a problem that needs solving. The real problem behind "needing a new car" might be a number of things: concern over your car's safety, failure of your car to pass inspection, lack of reliable transportation, or self-consciousness about being seen in a junker. Any of these possible underlying problems might have spawned the "I need a new car" solution, but please recognize that buying a new car is not the only strategy for resolving any of the underlying problems above, and it may not be the best strategy either. For example, carpooling, bike riding, motorcycle riding, leasing a car, fixing the junker, taking the city bus, or buying a more reliable (but not new) car may be better strategies if lack of reliable transportation is *really* the problem you're dealing with. But if your real problem is concern over your social status—whether peers think you're with-it—then different strategies such as stylish clothes, fancy jewelry, or an awesome sound system may get you closer to where you want to be. These alternative strategies also may be less expensive and easier to implement than your initial conclusion that "I need a new car."[1]

"Clubbing baby seals must be stopped" is another example of a preconceived solution posing as a problem. Think hard: What's the real problem with clubbing baby seals? What about it bothers you? What would you like to be different? Would you like the harvest to be less cruel? less gruesome and bloody? Is your concern really about preservation of a species? Is the real issue your feeling that it's wrong to make fur coats out of cute baby animals? The point made above cannot be overemphasized:

---

1. Note that if unreliable transportation really was the problem, certain strategies such as fancy clothes and jewelry would be completely ineffective. Similarly, "using public transportation" would be completely ineffective at solving the problem of low social status. The quality of a strategy depends entirely on the nature of the real problem.

To find a strategy (solution) that works—one that makes you feel that the problem has been taken care of—you first *must* identify exactly what you'd like to be different. If you bypass this most critical of problem-solving steps, be forewarned: you'll spend lots of time pushing ineffective "solutions."

Let's look at the baby seal situation again and see where different perceptions of the problem might take you. If the real problem in your mind is cruelty, then you can relax a bit, for clubbing the baby seals is a quicker, more efficient, more effective, and less painful method of harvest than the other common harvest methods of trapping and poisoning. If the real problem for you is having baby animals murdered in a bloody, gory way, then the way to make this situation more to your liking is for fur traders to poison the baby seals—it's bloodless and less gory, even if it's more cruel. If the real issue for you is health of the species as a whole, then a good strategy might be to double the annual harvest of baby seals so that there's more food for those that remain.

Playing out these scenarios, where different parts of an undesirable situation are identified as the potentially real problem, hopefully helps you see how incredibly important it is to be clear about what you want to change.

There are no shortcuts—if you're unwilling to invest the time and mental energy to be clear and specific about exactly what part of a situation you'd like to change, forget about it. It's better to do nothing than to jump on a misguided bandwagon.

Let's examine another preconceived solution that's posing as a problem: "We need to stop the cutting of tropical rainforest!" Once again, this call for action is a fuzzy solution in disguise. What's the real issue/concern/challenge/problem behind this preconceived solution? Is it loss of rainforest species? Which species? All species? Mosquitoes, AIDS, malaria, and smallpox too? Is the real problem loss of human cultures? Is the real problem excess CO_2 production and global warming?

It's easy to internalize a heartfelt battle cry and be intellectually lazy—to claim that the real problem is all of these things. Unfortunately, that kind of effortless thinking gets you nowhere.

To make headway on the concerns and issues that really matter to you, you first must sort through your feelings and reactions and find the smoking gun behind those feelings and reactions. Only then can you be effective at changing the situation so it's more to your liking.

How to Identify the Real Problem (the thing that you want changed)

As a first step to finding the underlying issue, challenge, opportunity, or concern behind a perceived problem, begin by asking two future-oriented questions:

- How would the situation need to be different for you to feel that you have successfully solved the problem?
- What would you need to see or feel or experience to feel good—to feel that the problem is no longer a problem?

Sometimes it's not all that easy to single out the exact part of a situation that's bothering you. Whenever you're spinning your wheels, feeling overwhelmed, or feeling unsure of the real underlying issue, try the following techniques: talking it out, the "repeat why" technique, exploratory writing, reality listing, and walking in other people's shoes. All five techniques are effective at helping you figure out which exact part of a situation you need to focus on.

Talking It Out

Talking it out is similar to what many of us do naturally when something's bothering us: We chew around the situation with others. You can steer these cathartic sessions into more productive venues by treating them as quests to find underlying realities. A good way to start down this path is to ask yourself and those around you, "Something's bothering me but I'm not exactly sure what it is. Here's the situation as I see it . . ."

After carrying on for a while about whatever comes to mind, start directing your thought processes a bit more. Begin by doing your best to state clearly what you think the problem is, then ask yourself and the others: "Is _____ the real issue? If _____ changes to _____, will I feel that the situation has been taken care of?"

Hashing things out is productive if you listen and process what's said, but it's not very productive if it's a whining competition. Work at listening, hearing, and processing what you and others have to say. Summarize and record key points on paper every few minutes so that insights aren't lost.

"Talking it out" certainly is a socially engaging way to search for the real, underlying issue, but beware! When you find that members of your group are all saying the same thing, don't accept that as proof positive that you've found the real underlying issue. Be cautious and curmudgeonly—play devil's advocate. Rather than embracing the party line, reverse positions and challenge the consensus view. Assume that the consensus view is wrong. Often times it is. Be a skeptic. Need convincing.

The "Repeat Why" Technique

The "repeat why" technique is very effective at helping you clarify the *real* problem because the technique forces you to look at underlying causes of the problem. The technique is very helpful at uncovering hidden assumptions.

Exposing hidden assumptions and problems early on is just what you want, but revelations can be very discouraging if you have been thinking that you are making great headway in the problem-solving process. Going back to the drawing board is never pleasant, but take heart: if you feel you are losing rather than gaining ground, consider the dastardly consequences of spending lots of time and money solving the wrong problem!

To illustrate how the repeat why technique works, let's say that your initial problem definition was, "We are not practicing sustainable management at the Franklin Forest." To help you assess if this problem definition is on target, ask your first *why* question: "*why* do we want to practice sustainable forest management?" Possible answers to this first *why* question might include:
• because Franklin Forest is the last piece of intact forest in the region,
• because ecological sustainability is our mission,
• because we need to maintain viable populations of wildlife,
• because our supporters expect us to.

For convenience of discussion, let's say that the first answer was the most reasonable response. Now ask the second *why* question: "Why is Franklin Forest the last piece of intact forest in the region?" Possible answers to this second *why* question might include:

- because other land is being developed without adequate land-use planning,
- because people don't care,
- because public forests don't generate taxes.

Going again with the first response, now ask a third *why:* "Why is land being developed without adequate land-use planning?" Possible answers to this third *why* question might include:

- because there are no zoning regulations or restrictions,
- because there is little enforcement of existing laws.

Going again with the first response, now ask a fourth *why:* "Why are there no zoning regulations or restrictions?" Possible answers to this fourth *why* question might include:

- because no one has introduced legislation,
- because local folks don't like the government telling them what they can't do.

Going again with the first response, now ask a fifth *why:* "Why hasn't anyone introduced legislation?" Possible answers to this fifth *why* question might include:

- because no one has thought to do it,
- because it would never pass,
- because we don't have the right kind of legislator.

As you can see, the repeat why technique opens up the problem so that you are forced to look at it in very different ways. This often helps you discern symptoms or disguised solutions that may be posing as problems. For example (working on the Franklin Forest sustainability problem), the repeat why technique might help you realize that the real issue (problem) is lack of regional planning, not sustainability of the Franklin Forest *per se.* This realization may prompt you to redefine the problem (e.g., "Lack of regional land-use planning is resulting in haphazard protection of our natural resources"), which in turn would redirect your problem-solving efforts and possible solutions. Finding the real problem permits you to focus energy where it is most needed.

The repeat why technique effectively and progressively exposes underlying issues, but be aware that, after several rounds of pursuing the *why?* behind a perceived problem, the exposed underlying issues begin to focus on social and cultural underpinnings and the like. For example, if we were to pursue a sixth *why* question above (e.g., "Why hasn't anyone thought to do it?"), the answers might include:

• because people are lazy,
• because our capitalist system doesn't reward such activities,
• because key players don't have much free time after taking their kids to soccer and violin practice.

These reasons, while insightful, may take you further into social-political-philosophical issues than you wish to go.[2] When you reach the point of irrelevance, backtrack to the level of *why* questioning that is most helpful.

Exploratory Writing

The purpose of exploratory writing is to expose your innermost thoughts and uncertainties and to foster new thoughts and connections. Recording them on paper enables you to study them carefully and dispassionately. It's therefore important to remember that exploratory writing is not about winning Pulitzer Prizes, it's about figuring out what's going on inside your head and putting pieces together. Two exploratory writing techniques are especially effective: freewriting and clustering.

Freewriting

Freewriting is what happens when you write in a journal or diary, or when you dash off a letter to a very close friend. In almost a stream of consciousness, your thoughts spill out on paper without concern over grammar, spelling, punctuation, syntax, or anything that high school teachers care about. Freewriting is sort of a written version of talking it out.

To use freewriting as a tool to find the exact parts of a situation that are unacceptable to you, just start writing about the situation. Splash anything and everything that comes to mind on a sheet of paper. Don't

2. A seventh *why* question could lead you even further into existential exploration (e.g., people are lazy because of their genetics, or because they didn't have a meaningful relationship with an authority figure, etc.).

struggle to write well, don't worry about what others will think of your writing, don't edit what you've written. Instead, let your guard down and let the words and ideas flow. Don't hold back—record any and all thoughts that float through your mind, whether they seem relevant or not—just do it. Focus solely on getting lots of words, thoughts, and ideas on paper in a hurry.[3]

If new thoughts are coming pretty slowly after ten minutes of freewriting, call it done. If words are still jumping on your paper, keep going until you start to bog down. After a short break, read your freewrite, *aloud.* Has your impression of the real, underlying issue changed at all? Probably so!

Here's an example of how freewriting was used to get at the heart of a problem with a coworker:

> George is holier than thou when it comes to being an "environmentalist" and it drives me crazy. And his "environmentalism" is phony—he drives an SUV and lives twenty-five miles away in a cabin in the woods. Sure, he's a crusader about recycling, but where does he get off having a four-wheel-drive, gas-guzzling car and driving fifty miles a day just to get to work? I asked him about it, and he said he needed a four wheel drive to get in and out of the driveway to his cabin; and he says he'd go out of his mind if he lived in the "burbs." It's clear that living in "the burbs"—in his mind—is about as low as you can go. Well, I grew up in the suburbs and, although it's not wilderness, we did have a big yard to play in, and lots of friends within walking distance. My parents moved there because the schools were better than in the city, and they thought it was a safer place to live. We lived in a raised ranch, just like a hundred other families, but it's the people, not the building that makes a place home. He's very judgmental about how others are living their lives and thinks his way is THE ONLY WAY. He really bugs me, and it's getting harder and harder to work with him because I don't respect him. I think he's a hypocrite.

That's the person's freewrite. With his thoughts and anger and frustration on paper, the freewriter now is able to identify some of the exact

---

3. A tip: If you're one of the many souls out there who has a hard time getting started when you write, try this trick: Put your pen on the paper and start writing "I can't think of anything to write, I can't think of anything to write." Continue writing this stupid sentence, without stopping, until your brain shifts its attention to the challenge at hand. Amazingly, this trick always seems to work—but only if you keep your pen physically moving!

parts of the situation that he'd like to be different. Here are the ones he identified:

- George has a holier-than-thou attitude.
- George's commitment to environmental issues is selective and self-serving.
- He's hypocritical.
- I'm really worked up about this but I don't know what to do about it.
- It's hard to work with him because he's quick to criticize the shortcomings of others but he's unable to see that he's not perfect either.
- He's very judgmental.
- He sees the world in black and white terms and is unable or unwilling to consider other points of view. For example, in his mind, suburban living is bad and country living is good. Period.

Those are the parts of the situation that the writer would like to make different. His next step would be to identify the exact desired outcome that he'd like for each.[4]

Here is an example of how freewriting was used to get to the heart of a river pollution problem:

> What's happened to the Pleasant River makes me want to cry. Doesn't anyone care? Twenty-five years ago the River was clear and we'd swim in it on hot days. We had so much fun—we built rafts, we fished for bass, we had pow-wows and built teepees and campfires. None of that is possible anymore—the land is all posted keep out! no trespassing! by people who have invaded with their big money. It's not right—they do what they want and think they own the river. They let their horses and fancy cattle and llamas and other weird animals do what they want in the river . . . no chance anyone would want to go swimming downstream of defecating cows! And the riverbanks on the downstream edge of their property are eroding like never before—mud, mud, mud! That's what's making the river so brown. It doesn't seem right that, just because they have money, these people can come in and do as they wish. How about the people who have lived

---

4. You may have noticed that a couple of the stated issues (George's selective commitment to environmental issues, George's holier-than-thou attitude) are assumptions the writer is making—not undeniable truths. It could be, in fact, that George really is very committed to improving the quality of the environment; it also could be that his brand of environmentalism is less selective and self-serving than the writer gives him credit for. If he is, in fact, assuming things that are not true, the problem lies with him, not with George. We'll show you how to separate assumption from reality when we introduce reality listing.

here all their lives? What about them? How about those who live or play downstream?

That's the person's freewrite. With her thoughts and anger and frustration on paper, she now is able to identify some of the exact parts of the river situation that she'd like to be different. Here are the ones she identified:
• Local people do not have access to the river the way they used to.
• The river is muddier than it used to be.
• Animal feces are making the river an undesirable place to swim.
• The new residents seem to do whatever they like, with little or no concern for others.
Those are the parts of the river situation that the author would like to make different. Her next step in solving this multidimensional problem is to identify the exact desired outcome that she'd like for each.

Clustering

Clustering is another exploratory writing technique that works well at exposing your inner thinking. It also draws out new thoughts and new connections. To begin clustering, turn a sheet of paper sideways and write down in the middle of the sheet a few words that describe what you think your problem really is about. Draw a circle around this description.

Now write down (in words or short phrases) whatever comes to mind vis-à-vis your problem description. Circle each entry as you record it (fig. 5.1), making connections to the entry that triggered it (if there was one). As with freewriting, let yourself go and write down anything and everything that comes to mind. Don't edit and don't mull over what you've done.

After several minutes, you probably won't be able to think of much else to add. Don't stop clustering quite yet. Instead, keep your pencil physically moving a couple more minutes by retracing the cluster circles and connecting lines. Surprisingly, this mindless activity usually leads to a few more useful insights.

Reality Listing

Some less-than-perfect situations, such as not being presentable in the morning, are easily made better. That's because we've come to cultural

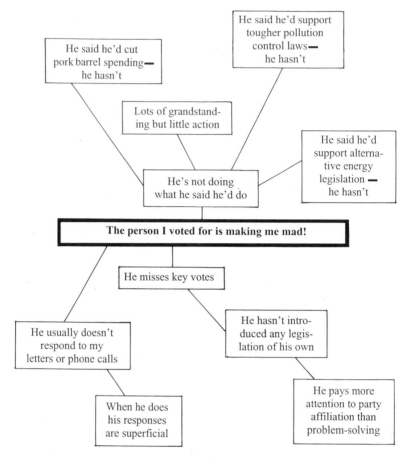

Fig. 5.1. Clustering
Example of how clustering was used to expose a disgruntled citizen's inner thoughts, concerns, and wonderings.
The problem: The citizen's elected representative is not representing her interests.

agreement on what's undesirable and problem solvers have developed strategies (e.g., hair brushes, deodorants, toothpaste) that consistently provide the desired outcomes. Environmental problems, and other life situations that keep you awake at night, are far more challenging than ratty hair or smelly armpits, because the exact problem may be unclear and/or because no prefabricated solutions are at your disposal. You're on your own.

Reality listing helps you dissect these types of complex problems into collections of smaller, more tractable challenges. By dividing and conquering, you can make headway on the most overwhelming environmental problems on earth. In fact, when tackling tough environmental problems, dividing and conquering is your only viable option. Single, all-encompassing solutions never work.

To use reality listing:

1. Make as complete a list as you can of the parts, realities, truths, and conditions of the situation you hope to change.

2. Now critically scrutinize each of these conditions/realities/truths, one at a time, asking, "Is this 'reality' unquestionably true, or could it be that you're just assuming it to be true?" Remove from your list all conditions/realities/truths that are merely assumptions. (Look carefully—assumptions arc hard to spot!)

3. Now critically evaluate each of the remaining conditions/realities/truths. Ask, "Is this one acceptable in its current form, or would you like it to be different?"

4. For each reality that you identify as not being acceptable in its current form, write down on paper a crystal clear statement of how the reality needs to change for you to feel good about it. These become your "desired outcomes," the collection of targets that (if hit) will move the situation from where it is to where you want it to be.

To illustrate, here's how reality listing could be used to tackle a vexing environmental problem: According to the electric companies, demand for electricity will exceed supply in the next ten years if new power plants are not constructed.

Step 1: List all the parts, realities, truths, and conditions of the situation that you'd like to change:
- a. I do not trust what the electric power companies are telling us.
- b. Demand for electricity will exceed supply within the next ten years.
- c. The appetite for electricity is much greater in the United States than it is in other countries.
- d. Most people don't care one way or the other.

- e. People are reluctant to change their energy-using ways.
- f. New power plants will need to be constructed.

Step 2: Is each "reality" above unquestionably true, or could it be that you're just assuming it to be true?
- I guess I'm just assuming that people don't care (d). Also, I am pretty sure that c is true but I'm not really sure (I need to check on that).

Step 3: Of the "realities" that are true, which ones would you really like to be different?
- The realities that I most would like to change are b, e, and f.

Step 4: What are your desired outcomes for b, e, and f?
- My desired outcome for b: The supply of electricity over the next twenty years will be adequate to meet nonfrivolous demands.
- My desired outcome for e: By the year 2010, per capita electricity use has declined by 15 percent.
- My desired outcome for f: No new major power plants will be constructed in the next fifteen years.

 The desired outcomes above become your targets or minigoals—what you need to achieve to feel that you've solved the electricity problem. (By dividing the big electricity problem into smaller, more focused subproblems and minigoals, your problem-solving task is much simplified).[5]

Walking in Other People's Shoes

To get a grip on reality and to see your "problem" as others do, always try walking in the shoes of someone else (the more different pairs of shoes you try walking in, the better). You'll be surprised at how different the situation begins to appear. Begin by adopting the persona of a hard-core conservative. How would a conservative see the situation? What would s/he want to change? In the conservative's mind, what would constitute an acceptable outcome?

---

5. Note that global and countrywide problems span diverse geographies, climates, economies, and cultures and "one size fits all" solutions are unrealistic. Broadscale problems remain virtually unsolvable until they are broken down and addressed at more relevant local levels. Heed Amory Lovins's exhortation, "Think globally, act locally!" It's more than a catchy expression.

Now assume the persona of a hard-core liberal. How would a liberal see the same situation? What would s/he think needs to change? What would be the desired outcome?

Now climb into the shoes of a religious fundamentalist and ask the same question, and do likewise for an atheist, an urban socialite, a farmer, an environmentalist, a developer, an inner city single mom, and a yacht club blue blood. Record on paper how each persona would view your problem and record the different perspectives on what would constitute an acceptable outcome.

Having looked at the situation through multiple lenses, does it look any different? If so, make appropriate adjustments in how you define the problem.

Some Wise Closing Thoughts

- Always define a problem as a complete sentence.
- Seek to define each environmental problem as a neutral description of the situation as it currently stands.
- Remember that *how* you define an environmental problem guides the rest of the problem-solving effort, including the "solutions" that you ultimately implement. Even the cleverest of solutions is worthless if it doesn't directly address the part of the situation that you want to change.
- The source of many environmental problem-solving impasses can be traced to people assuming they're working on the same problem when, in fact, they're not.
- Many environmental "problems" are as simple as others not being aware of (or not being concerned about) the situation that you think needs changing. Before leaping into a "we need to raise awareness!" crusade, however, ask yourself this: If you *are* successful at making people aware of the situation that troubles you, will that be enough? What if the newly aware people don't change their behaviors after being made aware of the situation—will you still feel that the problem has been solved to your satisfaction?
- Avoid searching for "solutions" until you know exactly what your desired outcome is.
- Always break environmental problems into smaller, more manageable

parts. Tackle the parts that you'd most like to change, one at a time. Don't trust mega-solutions; they never get you where you want to go.

• Treating preconceived solutions as "problems" gets you nowhere except into trouble.

Exercises

Provide a thoughtful, written response to each part of the following six challenges.

1. (Revisiting challenge 5.3): Fur traders head to the arctic each year to club baby seals to death for their fur. What can we do to stop this gruesome and cruel practice?

Your group is looking for solutions to this problem—what should be done?

a. Write down your exact desired outcome in a single, complete sentence.

b. Write down a clear, direct statement of the exact situation that is not to your liking—that is, the real problem. Present this also as a complete sentence.

c. Write down a viable solution that's likely to give you the outcome you want, and less likely to give you outcomes that you don't want.

d. Write down several plausible outcomes of implementing your newly proposed "solution," forecasting what might occur so you can see how well your strategy is likely to work.

How satisfactory (to you) are these outcomes? Almost certainly you forecasted some outcomes that were great and some that fell short of what you'd like. That's to be expected, for no solution is perfect. The issue therefore isn't *if* undesirable outcomes could possibly occur, it's what exactly *are* the undesirable outcomes? How troublesome would they be? How likely are they? What can be done to head them off at the pass?

2. Describe a global environmental situation that has been bothering you of late—something that you'd like to be different.

a. What is your exact desired outcome (goal) for this situation?

b. What is the *real* problem?

c. What might be a viable solution?

    d. What are the plausible outcomes if you were to implement this "solution"?

3. Describe a local or regional environmental situation that has been bothering you of late—something that you'd like to be different.
    a. What is your exact desired outcome (goal) for this situation?
    b. What is the *real* problem?
    c. What might be a viable solution?
    d. What are the plausible outcomes if you were to implement this solution?

4. Describe a recent problem situation where you fell into the trap of seeking solutions *before* figuring out the essence of the real problem.

5. Describe three preconceived solutions that people commonly think of or present as environmental "problems."

6. Use the "repeat why" technique to figure out the *real* reason why a particular environmental policy (or lack of policy) is not to your liking.

7. Use the "freewriting" technique to articulate why the following situation bothers you: Some of your actions and decisions contradict what you want to believe and stand for.

8. Use the "clustering" technique to get to the bottom of "the overpopulation problem" (as you see it).

9. Corporate greed and resource exploitation really bothers you, but you're not really sure what to do about it. Use the "reality listing" technique to identify exactly what needs to change for you to feel that this no longer is a problem.

10. Here's the situation you're wrestling with: whether it's wrong to buy and drive a big SUV, knowing that there are more efficient, less polluting forms of transportation out there. Use the "reality checking" technique to ferret out the real issues that are troubling you.

11. Return to page 1 of this book and reread your written description of the environmental problem that you'd like to see solved. Use a combination of problem definition techniques to ferret out (a) what you and your teammates agree upon as the *real* problem and (b) how the situation would

need to change for you to feel that your defined problem had been solved. State these clearly and succinctly on the lines below.

    a. What is the *REAL* problem? (Be sure to state it as a complete sentence.)

_____

_____

    b. What is the desired outcome?

_____

_____

_____

Chapter 6

The Yellow Brick Road: The "O" and "C" in DOC'S KEY

The preceding chapter showed you ways to ferret out and define "D," the *real* problem and desired outcome. This chapter will show you how to get the most out of the second and third problem-solving steps—objectives and constraints. These are the "O" and the "C" in DOC'S KEY to problem-solving success.

To review, the most important step in environmental problem solving is being crystal clear about the exact thing that you'd like to be different. As we have seen, this first step in the problem-solving road map is inextricably linked to accurately identifying the *goal*—how the situation needs to change for you to feel that the problem has been solved.[1] The next obvious step (once you know where you want to go) is finding a way to get there. That's where strategies (the "S" in DOC'S KEY) come in (chapter 7). But we're getting ahead of ourselves.

Objectives (the "O" in DOC'S KEY)

Objectives as Problem-Solving Yardsticks

Attaining your goal (i.e., solving your problem) may be the ultimate measure of problem-solving success, but solving an environmental problem

---

1. Many people confuse the terms "goal" and "objective" or use them interchangeably. That's unfortunate, because it can make following a problem-solving road map a little confusing. (That's why we have avoided using these terms until now.) To trained problem solvers, "goal" refers to what we have been describing as "desired outcome" in earlier chapters—the ultimate endpoint that constitutes problem-solving success (or, to say it another way, how a situation needs to be different for you to feel that the problem has been solved). "Objective," on the other hand, refers to a mini-goal—a manageable, intermediate outcome or target that gives you something tangible to shoot for. Whereas goals can be lofty or pie-in-the-sky (e.g., end world hunger, restore all wetlands to their precolonial state), objectives are always down-to-earth.

rarely happens in one fell swoop. Instead, environmental problems are almost always solved incrementally by meeting a string of mini-goals. These mini-goals are called "objectives."

Difficult environmental problems are solved by meeting objectives, not by searching for single, all-encompassing mega-solutions. Objectives dissect lofty goals into practical, specific, short-term targets.[2]

Some problem solvers describe goals and objectives in terms of a scavenger hunt: The goal is to complete or win the hunt; the objectives are the small successes that get you there. You can't reach your goal any other way.

To illustrate how objectives fit into the problem-solving road map, let's say that you've been asked to host a national conference for environmental leaders. This challenge may strike you as an exciting opportunity rather than a problem, until the glow wears off and you realize that much needs to happen to advance the conference from where it is now (nowhere) to where you want it to be (a resounding success). Your recognition that there's a gap between your goal (a successful conference) and your current situation (ground zero) launches you into problem-solving mode.

Let's say that, after working through the first problem-solving step (defining the problem—the "D" in DOC'S KEY), you identified three realities that are not at all encouraging:

• No one yet knows anything about the conference.
• The conference doesn't yet have a program or schedule.
• The participants don't yet have a place to meet, sleep, or eat.

Each of these constitutes a "problem" because each situation is different from what you'd like it to be.

With the problems in front of you, the goals become clear:

• for all relevant people to know about the conference,
• for the conference to have a program or schedule,
• for the participants to have a place to meet, sleep, and eat.

Good. You now know what it will take for you to feel that each of these three problems has been solved. That's an essential first step.

But now what? That's where "setting objectives" comes in: daunting goals are broken down into challenges that are more specific, tangible, and

---

2. As with goals, objectives tell you *where* to go, but they don't tell you *how* to get there. That's the province of strategies (the next chapter).

manageable. To illustrate, the first goal might be broken down into the following objectives:

- Finalize a decision by January 15 on where and when the conference will be held.
- Inform all environmental leaders of the upcoming conference by June 7.
- Send out registration materials within one week of receiving a request for information.

Possible objectives for the second goal might be:

- Draft a conference program for review by March 1.
- Receive feedback on the draft program by March 15.
- Finalize the conference program by April 15.

And possible objectives for the third goal might be:

- Decide on conference rates, meeting rooms, and lodging by May 1.
- Finalize banquet preparations by July 15.
- Deliver a sumptuous banquet at 6 P.M., August 23.

As you can see, breaking goals into specific objectives makes the conference problem much less overwhelming.

Dividing and Conquering

With a set of reasonable objectives on paper, step back for a moment and consider whether additional "dividing and conquering" would make your problem-solving life easier. Consider each of your objectives, one at a time: would it be easier to meet this objective, or delegate tasks, if the objective were broken into smaller, more manageable subobjectives?

For example, the objective, "to deliver a sumptuous banquet at 6 P.M., August 23," points you in the right direction by clarifying one part of your desired outcome. But delivering a sumptuous banquet is a far from trivial challenge; getting your mind around all of the moving parts at the same time might overwhelm you. To simplify this complicated objective and be more effective, divide and conquer this objective into more manageable subobjectives,[3] such as

- To finalize the banquet location by June 10,
- To serve hors d'oeuvres from 4 to 5:30 P.M.
- To have mashed potatoes on the banquet table by 6 P.M.

3. This process of breaking environmental problems into smaller, more tractable challenges should be familiar to you now, for dividing and conquering is a fundamental theme of effective problem-solving. Use this strategy often.

Each of these could now become an objective in its own right: you'd then only need to strategize ways to make each become a reality. For example, to have the mashed potatoes appear on time, you could assign specific tasks to others:

- *To helper 1:* "Go to the store today and buy fifty pounds of potatoes. Have them back here by noon."
- *To helper 2:* "By 4 P.M., have forty pounds of good potatoes peeled and sliced."
- *To helper 3:* "Have the forty pounds of sliced potatoes cooked and mashed with milk and butter by 6 P.M."[4]

In the end, if the tasks are completed as indicated, you'll have conquered one objective and one piece of the banquet "problem." If you're able to meet all of your other subobjectives for the banquet, you will have met your objective for the banquet as a whole (to deliver a sumptuous banquet at 6 P.M., August 23). You also will have met a major part of your overarching goal—to host a great conference.

Here's another illustration of how you could divide a challenging objective into smaller, more manageable subobjectives. Let's say that one of your objectives is to deliver environmental education programs to inner city children by September 15, 200X. With this (seemingly reasonable) objective identified, the next logical step would be to look for strategies that might get you where you want to go. As you start searching for ways to meet this objective, however, it quickly dawns on you that there are lots of very relevant and daunting considerations that you haven't yet addressed: What inner city children should you target? How will you find them? Once you find them, will they be interested in participating? When will you offer the programs? Where? What environmental programs will you offer? If there are expenses, who will pay for them? Should you be concerned about liability?

These are very different challenges, and no single strategy can give you the answers that you need. It therefore makes sense to break the original objective into smaller, more manageable subobjectives. For example:

- to decide, by June 15, the student group that you wish to target;
- to summarize on paper, by June 30, your vision for the inner city educational initiative;

---

4. Notice that each objective is a desired endpoint, not a step-by-step explanation of how to get to the endpoint.

• to establish, by July 30, working relationships with at least two people who are able and willing to help link you to possible student groups.

Tackling each of these subobjectives one at a time enables you to generate strategies that will target the exact result that you seek. In so doing, you'll more easily reach your final desired destination.

When an environmental problem overwhelms you, or when a goal seems impossibly difficult to achieve, break the problem and goal into smaller pieces and treat each piece as a free-standing problem and goal.

You then can craft very targeted objectives for each subproblem/subgoal and, in so doing, make solving the big problem infinitely easier.

How to Know Useful and Useless Objectives When You See Them

Useful objectives are "S-M-A-R-T." They are:

Specific

Measurable

Attainable

Reasonable, and they identify a

Time by which the objective needs to be met.

Here are some examples of objectives that are useful:

• Reduce CO_2 emissions from all municipal buildings by 10 percent by June 15.

• Raise $40,000 by October 10 for our environmental cause.

• Increase membership in our organization by 12 percent by New Year's Day.

• Restore at least twelve colonies of rare species "X" in Pelican Marsh by June 15, 2010.

• By August 1, inform 20 percent of Clinton County landowners about rules and regulations contained in the new "Logging Practices Act."

   Here are some examples of objectives that are useless:

• Reduce CO_2 emissions.

• Raise money for our environmental cause.

• Increase membership in our organization.

• Inform landowners about the new "Logging Practices Act."

Note that each *useful* objective above is **S**pecific, **M**easurable, **A**ttainable, **R**easonable, and has a **T**imeline for completion. Note also that each *useless* objective lacks these qualities.

How to Craft Useful Objectives

Since objectives are the specific, desired (intermediate) outcomes of your problem-solving effort that will collectively take you to your goal, the easiest way to initiate a search for suitable objectives is to examine your defined problem and goal carefully to see what you really want. First ask, "What collection of objectives, if achieved, will get me to my goal (and cause me to believe that the situation is no longer a problem)?" Record your answers to this question so you don't lose them. Be as specific and comprehensive as you can.

Now review each of your recorded objectives critically. Check to be sure that:
- each objective includes a verb (verbs force you to think effortfully about what you want as outcomes);
- each objective focuses on a single specific, desired, intermediate outcome (this breaks the problem and goal into manageable pieces);
- each objective is measurable (how else can you know if a "solution" has worked?);
- each objective is attainable (if objectives cannot be met, they are not useful targets);
- each objective is reasonable (considering time and resources at your disposal, and your wish to meet other objectives, will you really be able to meet this objective?);
- the target date for meeting the objective is clearly stated and reasonable;
- you will feel that your goal has been reached if you satisfy all of the stated objectives;
- each objective fits the problem and goal as you have defined them (if it doesn't, rethink your objective, your defined problem, and defined goal— they must be obvious extensions of one another);

When crafting objectives, plan on generating several (3–5) for each problem or subproblem. If you identify more objectives than that, simplify the problem-solving effort by dividing your defined problem into smaller, more specific and manageable subproblems. You then will be able to set objectives for each subproblem, solve those subproblems, and in so doing, solve the bigger problem.[5]

Goals, Objectives, and "Modeling the Ideal and Working Backwards"

As stated earlier, many people, including some environmental professionals, use "goal" and "objective" interchangeably. To trained problem solvers, the two terms are quite different. "Goal" summarizes what a situation will look like when the problem has been completely solved (e.g., "for navigable waters in the state to be free of nonnative species"). "Objectives," in contrast, are the specific, piecemeal outcomes that need to occur to reach your ultimate goal (e.g., "to eradicate nonnative species X from Joe's Pond by September 15").

Solving environmental problems by meeting objectives is a strategic planning approach called "modeling the ideal and working backwards." This technique, which emphasizes deciding what you want (objectives) and then fashioning ways to achieve it (strategies to meet your objectives), works especially well for environmental problems.

Sometimes, even after lots of work, it's hard to pinpoint what your specific objectives really need to be. When this occurs, it usually means that you have not zeroed in on the *real* problem or the *real* goal. Try rethinking your problem definition and goal by sketching pictures, diagrams, and graphs of your desired outcomes. Even the most primitive "dummy" pictures, diagrams, and graphs will help you see your problem and desired outcomes more clearly.

In summary, clear objectives serve two critical functions in environmental problem solving: (1) they provide specific problem-solving targets, and (2) they serve as measuring sticks against which possible strategies can be evaluated. As you meet each objective, you solve a known percentage of the overall problem. You're therefore able to monitor, and report to others, your progress in solving the overall problem.

5. Many fledgling problem solvers balk at dividing a problem into subproblems (and accompanying subgoals and objectives) because, at first glance, it seems to make things *more* complicated not *less*. Try to think beyond this first glance.

Constraints (the "C" in DOC'S KEY)

Boundaries, Limitations, and Hidden Assumptions

Trying to solve important environmental problems is never far from crisis management, so it's easy to think that there's never enough time or money to do things right. A more accurate assessment would be that there's never enough time or money to do things wrong.

Passing over "constraints" (i.e., problem boundaries, limitations, and hidden assumptions) is one common but very dangerous mistake made in environmental problem solving. Zipping past this step in the problem-solving road map may save you a little time in the short run, but the consequences always catch up with you later, often in disastrous ways.

"Constraints" are the real and imagined boundaries that circumscribe how you go about trying to solve a problem.

When you clarify the true boundaries, limitations, and constraints of an environmental problem, you place the problem in context. People outside your problem-solving group then can see where you're coming from, where you're going, and where you're not going. If you've overlooked something important, they can alert you to what that "something" is, before it's too late.

Clarifying the boundaries of an environmental problem also helps you focus energy where it's needed most. If you are tackling a problem related to "sustainability," for example, you need to be clear about details such as what exactly you wish to sustain, who will do the sustaining, and how much time and money you can expend on the sustainability effort. Similarly, if you're working on a problem related to urban sprawl, you need to define clearly what you mean by "urban," and what you mean by "sprawl"; you also need to be clear about other boundaries, such as who should be included in the process and who should not. When you find your problem-solving effort wandering aimlessly, it's usually the result of your not having defined the problem's boundaries as clearly as you should.

Making decisions about your problem's boundaries can be challenging, especially when you're faced with seemingly arbitrary decisions about what is part of the problem and what's not. But decisions about the scope of an environmental problem must be made sometime, and sooner is always easier than later.

Hidden Assumptions and "Truths"

Delineating the boundaries of an environmental problem makes solving the problem easier, partly because it flushes into the open hidden and faulty assumptions. Faulty and hidden assumptions are the bane of problem solving. The only way to neutralize them is to find them before they lead you astray.

Fig. 6.1 is a teaser to see how well you spot hidden assumptions: without lifting your pencil from the paper, connect the dots on page 86 with no more than four straight lines. (If you have seen this puzzle before and know "the answer," you're not off the hook: approach the puzzle anew and find three new, different answers).

Were you able to solve the dot problem? If you were, congratulations! Go back and find two entirely different strategies.

If you had trouble solving the dot problem, it's because your strategy-generating creativity was constrained by unnecessary or faulty assumptions.[6]

Unnecessary or wrongful assumptions compromise your ability to solve environmental problems effectively, but identifying these hidden boundaries (which run rampant), is not so easy. Consider, for example, the following:

- "Natural medicinal plant remedies are safer than pharmaceutical drugs made from chemicals."
- "You can't put a price on human life."

Many people assume these two statements are true and act accordingly. But are they *really* true? Or are they assumptions or deeply held convictions rather than scientific realities?[7] Not recognizing the difference spells trouble, for when convictions or assumptions are misconstrued as "truths," bad solutions usually result.

As elaborated in another chapter, each of us holds hidden assumptions

6. When you find yourself immobilized by a problem that you think should be easy, it is likely that at least part of your stymied state is caused by self-imposed (unnecessary) boundaries.

7. Many scientists, in fact, would say that botanical extracts are less safe than controlled chemical drugs, and many economists would offer the reminder that you place a value on life each time you decide how much you're willing to spend to make your home, road, and car safer. Would you spend a dollar to retrofit your car so it was impossible for you to be killed in an accident? Probably. Would you commit 99 percent of your life's earnings to do the same thing? Probably not.

Fig. 6.1. The Dot Puzzle
The Challenge: Without lifting your pencil from the paper, connect the dots with no more than four straight lines.

and "truths" that were spawned by our upbringing, culture, religion, and socioeconomic status. Differences of perspective in problem solving are valuable because they unmask preconceptions, and they reveal hidden assumptions and unnecessary boundaries. Once identified and exposed, preconceptions then can be discussed and evaluated in the open, before they create massive problems of their own.

Preconceived notions and assumptions are the most common forms of unnecessary, self-imposed boundaries. You can remove at least some of these by scrutinizing your problem, objectives, and boundaries carefully. Try to shift your thinking from, "I didn't know I could do that" to "It didn't say I couldn't do that!" With that new thought pattern in mind, try the dot problem again and see if you have better luck. Begin by reading the instructions again, *but this time read the instructions carefully!* Critically evaluate *exactly* what the dot instructions say you *can* do, and exactly what the instructions say you *cannot* do. Let the instructions speak for themselves— *assume nothing!*

You now should be able to generate a number of strategies to the dot problem, because you have discarded faulty assumptions that acted as solution blocks. For example, you may now realize that nowhere do the directions state that you cannot extend lines into the outward surrounding blank space, or that you cannot use nonstraight lines, or that the paper must remain unfolded, or that a regular, thin-leaded pencil must be used.[8] By shedding unnecessary boundaries such as these, your mind is liberated to think more freely and creatively and to generate more and better solutions. Given the inherent complexity of environmental problems, that type of mental liberation is crucial.

Techniques to Define Your Problem's Boundaries and Hidden Assumptions

Many of the techniques discussed in chapter 5 that help you ferret out the *real* problem (talking it out, "repeat why," exploratory writing, reality listing, walking in other people's shoes) work well for exploring problem boundaries. Troubleshooting (chapter 9) also works well. Before trying any of these techniques, however, first create a skeleton profile of your problem's boundaries and hidden assumptions by answering on paper the following questions. *Specifically:*

1. Who is part of the problem?

2. Who is not part of the problem?

3. Who needs to participate in the decision-making process?

4. Who should not participate in the decision-making process?

5. When must a strategy be available for implementation?

6. For what time period must the strategy be effective?

7. What financial constraints are relevant?

8. How much time do you really have to devote to solving the problem?

9. What are the physical boundaries of the problem (where does the problem begin and end)?

10. What legal considerations limit which problem resolutions are possible?

11. What political factors limit which problem resolutions are possible? (Think effortfully about this—don't take the easy or lazy way out!)

---

8. A few possible strategies are found on p. 94, but no peeking until you have given the problem a good try!

12. What environmental factors limit which problem resolutions are possible?

13. What unacceptable impacts limit which problem resolutions might be viable?

14. What other limitations, boundaries, and constraints define the scope of the problem?

15. What assumptions are you making?

Responding on paper to the fifteen questions above usually exposes a number of wrong or unnecessary assumptions. Revise your list of boundaries and assumptions as needed, and then use the other recommended techniques (mentioned above) to further define appropriate boundaries. Now critically evaluate with colleagues each of your revised boundary definitions by asking and answering, "Is this necessarily so?" and "Am I forgetting something?" Use insights from this critical analysis to revise and clearly state your boundary definitions.

As a final check on how accurately you defined the problem's boundaries, and how wantonly you fell prey to faulty assumptions, try running your problem and objectives by people outside your group. Sit back and listen carefully to how they would go about tackling the problem. Don't be surprised if they see through your self-imposed constraints and suggest possibilities that had not occurred to you.

Who? What? Where? When? Why? How? (The 5Ws and H Technique)

This technique is used by journalists to uncover important pieces of a story. Like the "repeat why" technique described in the last chapter, the "5Ws and H technique" opens up perceived problems so that you can see what's really going on. This technique is very effective at forcing you to re-examine preconceived notions.

To use this technique to help clarify your problem's constraints, boundaries, limitations, and hidden assumptions, provide answers for each of the 5 Ws and H. Using the Franklin Forest example from before, "We are not practicing sustainable management at the Franklin Forest," ask and answer:

• Who is "we"? (*Possible answers:* the problem-solving team, a board of directors, a consulting firm, the community, local government, the state Agency of Natural Resources, the forest manager, etc.)

- Who manages the forest?
- Who could manage the forest?
- Who should manage the forest?
   Similarly, ask the other "W" and "H" questions:
- What does "sustainable" mean? (*Possible answers:* provides enough income to offset expenses, maintains a consistent population of deer from year to year, produces as much wood each year as is harvested, etc.)
- What exactly constitutes the "Franklin Forest"? (*Possible answers:* the land defined by property boundaries, the plants and animals that reside permanently within the property boundaries, all biota that use the property some time of the year, acreage that has trees on it, etc.)
- What does "management" mean?
- What aspect of the current management is not sustainable?

Be as comprehensive as you can (ask as many "who" questions as you can think of, and do likewise for the other Ws and H), and be equally comprehensive in your responses to each question. Continue this question asking and answering until all 5Ws and H have been addressed. As with all techniques, record questions and associated responses on paper so they can be reviewed later as prompts to clarify your problem's constraints, boundaries, limitations, and assumptions.

Boundary Examination Technique

When you define an environmental problem on paper with words, you knowingly and unknowingly place boundaries on the problem via the words that you use. Selected words guide thought processes because words carry assumptions with them. That is desirable if you define a problem accurately but very undesirable if you do not. Faulty and unnecessary assumptions limit your ability to generate and implement creative, effective strategies.

The boundary examination technique dissects a problem definition into its component parts (words), and directs you to evaluate each word or cluster of words for assumptions and boundaries. This allows you to evaluate the validity of each identified assumption and boundary and select whichever assumptions and boundaries are appropriate.

To examine boundaries of a problem definition, carefully read through your problem statement, looking for words or clusters of words that may impose assumptions, boundaries, or ambiguities on how the problem is defined. Underline these words or clusters of words and list their assumptions and boundaries.

Now, for each assumption and boundary, list alternatives. This process is demonstrated in the following example: Let's say that you have defined your problem as, "*We* are not *managing* the *Franklin Forest* in a way that *generates sufficient revenue* to *pay taxes.*"

- *Assumptions and Boundaries: We* means the problem-solving group. *Alternatives: We* could mean the forest manager, the state, the community, an ad hoc group, the Forest Service, etc.
- *Assumptions and Boundaries: Managing* means that humans impose their will on the forest, or that they do something physical to the forest, or the forest will be changed from what it is now. *Alternatives: Managing* could mean letting the forest impose its will on humans, changing how we think about the forest, doing nothing whatsoever, letting nature manage itself without human interference, recreating a "natural," precolonial forest ecosystem and then leaving it alone, etc.
- *Assumptions and Boundaries: Franklin Forest* means trees, or the physical boundaries of the property. *Alternatives: Franklin Forest* could mean the entire forest ecosystem (land, soil, plants, animals, air, and people), the physical dimensions of the property, all surrounding landscapes that interact with it, the view-shed, etc.

After all underlined words in the problem definition have been similarly reconsidered, critically evaluate all of the underlined words and their assumptions, boundaries, and alternatives. Select the most accurate wordings for each and incorporate them into a modified problem definition that better reflects the true nature of the problem. In so doing, you clarify the problem's boundaries, constraints, and assumptions. Just as important, you clarify your goal and your objectives.

The Farmer's Overview

The first three steps of DOC's KEY to problem-solving success (problem and goal definition, objectives, constraints) are precursors to the rewarding part—finding strategies that will change the problem situation so that it's more to your liking. Until now, the problem-solving steps have been treated as independent entities, which of course they're not. Before heading to the next step ("Finding Strategies"), it's wise to examine your problem definition, goal, objectives, and constraints as a unified whole to be sure that they're in keeping with one another. As a test of how cohesive your definition, objectives, and constraints may be, work on developing a convincing "farmer's overview."[9] Never proceed further in trying to solve a problem until you have crafted a farmer's overview that's convincing to *you*.

The "farmer's overview" is a short, unambiguous, compelling summary of what your problem is all about, with enough context, background, and desired end products thrown in to make the overview interesting and meaningful. A good farmer's overview provides others with a quick explanation of the *real* problem you're tackling, why it's important, and what you think constitutes an acceptable outcome.[10]

To develop a useful farmer's overview, first imagine that a farmer, grandparent, friend, or other smart, thoughtful person has asked what you're working on. How would you describe your undertaking in a convincing, interesting way? Begin your crafting of an overview by simply talking to yourself aloud. If you can't think of anything to say, just make your mouth move and force words out—something useful will eventually surface. Tell yourself about the problem that you're working on and why it's important. Explain aloud how the situation needs to change for you to feel that you've solved the problem.

Once you've created an overview that's more or less convincing, record it on paper so that you have something to work with. Begin fine-tuning your farmer's overview by checking to see if it's more than half a page. If it is, it's too long—you haven't really figured out what you're doing or why. You don't yet have the bottom line.

With a few solo practice deliveries under your belt, try your farmer's overview on friends and colleagues. Pay careful attention to their reactions and revise your overview accordingly.

Now try your revised overview on some other unsuspecting souls, again making revisions as necessary. Continue this process of foisting revisions of your farmer's overview on acquaintances until both you and your listeners find your overview compelling and convincing.

Don't despair if your early efforts at farmer's overviews receive luke-

9. To stay in business, farmers need to be practical, resourceful, and clear-headed problem solvers. It's been my experience that they're quick to spot fuzzy or unrealistic thinking.

10. Many problem solvers find that the farmer's overview helps them see and remember the bigger picture. Sustaining focus and interest in tough problems is immensely important once the novelty of a problem wears off.

warm receptions. Crafting tight, accurate, meaningful, compelling overviews takes practice. If repeated efforts at improvement fail (i.e., the dazed look on your listeners persists), it probably means that you have not explained how your problem fits into the bigger scheme of things. Environmental problems often need to be placed in context to be meaningful or convincing, so try asking aloud—and answer!—the questions, "So what? Who cares?" Working on these questions and answers aloud forces you to identify weaknesses in your overview, and this invariably leads to increased clarity.

In summary, an effective farmer's overview is very much like an effective executive summary or sales pitch; in half a minute's time, a good farmer's overview explains the problem and its importance in language that anyone can understand. A good farmer's overview also articulates the desired outcomes and how they collectively constitute a solution to the problem. In short, it paints a clear picture of the problem and how and why you're trying to solve it. Use it as a litmus test for how accurately you have defined the real problem and desired outcome. Make adjustments as needed.

Inviting Stakeholders into the Process

With a convincing farmer's overview in hand—one that accurately mirrors your problem definition, goals, objectives, and constraints—it's time to seek reactions from interested outside parties (stakeholders). After presenting your integrated but bare-bones assessment of the problem situation (your farmer's overview), ask for critical feedback—and listen carefully to that feedback! If feedback is slow in coming, probe a bit by asking these questions:
- Have I captured the true essence of the problem as you see it?
- If we get the desired outcome (attain the goal that I presented), will you also feel that the problem has been rectified and is no longer an issue?
- Are the objectives reasonable?
- What have I forgotten?

Expect stakeholders to force into the open issues and concerns that you've overlooked; also expect them to pounce on fuzzy thinking. The feedback you receive may send you back to the drawing board—a discouraging outcome for sure—but receiving a reality check early on, when you can still fix what's broken, is far preferable to receiving the same message later, when it's too late.

Seeking input from stakeholders at this particular stage of problem solving is essential for another reason: stakeholders need to feel that they're part of the process before they'll buy into an outcome wholeheartedly. Often, the difference between success and failure lies in whether stakeholders have had their perspectives heard and taken seriously. Never underestimate the power of disgruntled stakeholders! When people feel left out of the process, they can create more difficulties than you're able to imagine.

Exercises

Divide and conquer the following two situations by setting forth three S-M-A-R-T objectives for each of the following perspectives, (a) and (b).

1. There is a proposal to extend the town's water and sewer lines to Freemont Heights. (a) You are leading the charge to have the proposal approved; (b) you are leading the charge to have the proposal defeated.

2. Congress will vote soon on a proposal to drill for oil in some national parks. (a) You have been hired by the oil industry to make sure that the proposal is approved; (b) you have been hired by an environmental coalition to make sure that the proposal is defeated.

Assume that exercises 3 and 4 represent situations that you see as problems. For each of these problems, clearly state (a) a reasonable goal; and (b) two S-M-A-R-T objectives.

3. Few people on your street are recycling.

4. Water from the Colorado River (which once flowed to Mexico and was used by farmers to irrigate their subsistence crops) is now diverted to water the lawns of rich Americans who choose to live in the arid Southwest.

5. Identify a recent political action that upsets you. For you to feel that this situation has been changed to your satisfaction, what would the situation need to look like (i.e., what would be your goal)? What might be your S-M-A-R-T objectives?

6. Identify a local environmental problem that you would like to see solved. For you to feel that this situation has been changed to your satisfaction, what would the situation need to look like (i.e., what would be your goal)? What might be your S-M-A-R-T objectives?

7. Identify a state or countrywide environmental problem that you would like to see solved. For you to feel that this situation has been changed to your satisfaction, what would the situation need to look like (i.e., what would be your goal)? What might be your S-M-A-R-T objectives?

8. Many environmental problem-solving efforts have failed because the "objective-setting" stage of problem solving was not taken seriously. Give two examples of such failings.

9. Describe a problem-solving effort that has been stymied by hidden or false assumptions.

10. Revisit the important environmental problem that you identified on page 1 of this book. Now turn to page 76 to revisit how you clarified your definition of this problem and associated goals. Using techniques described in this chapter, break down your defined problem and goals into (written) S-M-A-R-T objectives. Remember to "divide and conquer" your problem into smaller subproblems and subgoals if (a) you find yourself needing to generate more than half a dozen objectives to attain your goal; or (b) this part of the problem-solving process is seeming too overwhelming. Be sure, of course, to work hard at this. You will be revisiting these objectives when you move on to subsequent steps in DOC's KEY.

11. Revisit 10 above. Using techniques described in this chapter, define (on paper) your problem's boundaries, constraints, limitations, and hidden assumptions. Be sure, of course, to present and record them clearly so that they could be easily understood by others.

12. Revisit 11 above. Craft a good (written) "farmer's overview" for the environmental problem that you are tackling. As always, be sure that your product is clear and easy to find when you and others need to refer to it in the future.

Examples of strategies for the dot problem (p. 86):
• Use curved lines to connect the dots (it didn't say you couldn't).
• Use a 4″ wide paint brush and connect them all with one swath.
• Carefully fold the dot matrix so that the dots lie directly atop one another. Connect the dots by piercing each stack of dots with a needle.
• Cut out the dots and lay them on a table so they touch one another.

Chapter 7

How to Pull Ideas Out of Thin Air: The "S" in DOC'S KEY

The first three steps in DOC's problem-solving KEY—(1) singling out the *real* problem and goal (D), (2) setting S-M-A-R-T objectives (O), (3) identifying real and imagined constraints, boundaries, limitations, and assumptions (C)—clarify where you want to go. Knowing where you want to go doesn't tell you *how* to get there, however. That's the purview of idea generation (finding strategies—the "S" in DOC's KEY), and the subject of this chapter.

Obstacles to Creative Thinking

You've probably noticed that some people exude creativity in everything they do; others (perhaps you) wallow in unimaginative sameness. What makes some people imaginative and others not? Some of it is explained by genetics, of course, but much of it is explained by how you go about the creative process. Less ceative people:

• worry too much about what others might think; and
• evaluate the goodness or badness of ideas before the ideas have time to mature.

Maneuver around these two obstacles and you'll liberate creativity you never knew you had.

Easier said than done? Commit to the following:

• Surround yourself with supportive, positive-minded people who have a joyful, uninhibited approach to life.
• Stay away from nay-sayers.
• Believe that you are no dumber than the other bozos out there (you probably aren't).

- Believe that you can do—or at least try to do—anything.
- Experiment, be flexible, try new things.
- Be playful and carefree with ideas; adopt a "what-the-hell" attitude.
- Believe that wild and crazy ideas are every bit as valuable as sensible ideas (they are).
- Measure creative success by the *quantity* of ideas that you produce.
- Abandon daily ruts and routines.

Last, frequently remind yourself of the personality traits that you and others most value and respect; emulate those traits. Is it sameness? Is it trudging along mindlessly, trying to be a generic sheep like everyone else? Is it waiting for someone else to take the lead so that you can follow? Hopefully not! So get a grip:

Stop worrying about what others *might* think if you stray from the norm. Worry instead about what others *will* think if you're a boring drone who never leaves the beaten rut.

Shifting the focus of your worry works wonders at neutralizing the first major obstacle to creative thinking. The second major obstacle to creative thinking—judging an idea's value the moment it comes to mind—can also be neutralized if you remember that:

New ideas are rarely usable when first generated. Like new bottles of wine, ideas need to mature and develop character before becoming useful. This maturation process cannot happen if ideas are discarded prematurely.

People seem naturally inclined to judge and dismiss offbeat ideas because (1) they assume that an idea has no value if it's not immediately usable; (2) they fear that "useless" ideas, if not disposed of immediately, may slip through the cracks and wreak havoc later on; or (3) they worry about how they themselves will be judged if they take a path less traveled.

Separating the generation of ideas from the evaluation of ideas is essential because the two types of thinking (creative and critical) use very different and antagonistic parts of your brain. When you try to think creatively and critically at the same time, your brain waves clash and both types of thinking suffer—your creative thinking isn't all that creative and your evaluative (critical) thinking isn't all that sharp. After lots of mental struggling, you're no better off than when you started.

Following a problem-solving blueprint such as DOC's KEY helps you manage creative thinking and critical thinking so that you get the best from each. When you need ideas, you're able to relax, have fun, and be productive. When you need to judge the quality of your ideas and make decisions, you're able to think clearly and decisively. You need to be able to do these things well to solve environmental problems.

How to Find Ideas Where There Are None

Begin by:
1. deferring all judgment of ideas until another time, and

2. worrying less about what other people think.

Then:
3. Discuss your problem with people outside your circle.

Discussing Your Problem with People Outside Your Circle

Always talk over your problem with outsiders before choosing one idea over another. Sharing your thought processes openly with people from other walks of life forces you to be clear about your problem, your desired outcomes, and how you aim to get where you want to go. People outside your inner group have fresh minds and perspectives and are able see the problem in a different light than you do. This often leads to suites of very different ideas and strategies.

When you discuss your problem with someone outside your circle, two valuable outcomes are likely: (1) the listener will effortlessly suggest obvious, simple strategies that had not occurred to you, or (2) the listener's eyes will glaze over in stupefaction and disinterest. The "glazed-eye" reaction indicates that your thinking about the problem is still fuzzy and you need to work more on your farmer's overview.

Brainwriting

The brainwriting technique is a powerful tool used by think tanks to generate lots of different ideas in a hurry. To illustrate how brainwriting works, let's say that you're looking for ways (strategies) to meet the following S-M-A-R-T objective:

• "To raise $180,000 for our conservation fund by January 1."

Begin by assembling a group of three to ten people (six to eight is ideal). If your group is larger than ten, divide members into smaller subgroups of five to eight.

Now reframe your objective as a brainwriting challenge by asking "In what ways might we . . ." in front of the objective:

• "*In what ways might we* raise $180,000 for our conservation fund by January 1?"

Ask each brainwriting participant to write this challenge at the top of an otherwise blank sheet of paper. This reminds brainwriters to stay focused on the same challenge.

The actual brainwriting process now begins: Brainwriters are asked to start listing on their sheets of paper anything and everything that comes to mind related to the stated challenge. No editing or evaluating![1] The goal of brainwriting (as with other idea-generating techniques) is to generate as many thoughts, ideas, and strategies as possible. Brainwriters must believe that crazy ideas, sensible ideas, unusual ideas, and commonplace ideas are equally valuable. At this stage of problem solving, they are. Here's how a brainwriting session unfolds: After a minute or two of listing ideas on your sheet of paper, your outpouring of new thoughts will begin to wane—you'll find yourself working to come up with new ideas. When you notice this happening, ask a teammate to switch sheets with you by requesting a "Switch!" out loud. One of your teammates should immediately come to the rescue and switch sheets with you.

Now read through your teammate's list of unedited thoughts. Almost certainly some of them will prompt you to look at the brainwriting challenge in new ways, and this will trigger a new outpouring of thoughts, strategies, and ideas. Add these newly inspired ideas to the bottom of the

---

1. It might seem senseless to record wild and seemingly unrealistic ideas if the ideas have no chance of ever being implemented. But it's far from senseless, and here's why: wild, uncensored ideas are valuable early on in idea generation because they help you and others look at challenges in new, fresh ways.

list that you now hold. Continue adding new ideas to this list until your output of ideas once again begins to stall. As before, request a "Switch!" out loud and trade lists with a different teammate. As before, read through your newly acquired list of raw ideas and add new ideas to the bottom of the list as they come to you. Continue this process of recording ideas and switching lists for ten minutes.

To conclude a brainwriting session, the brainwriting facilitator asks participants to finish whatever they are writing, and then to take turns reading aloud a couple of interesting ideas from the brainwriting sheet each is holding. If any of these verbally shared ideas prompts you to think of something new, record the new thought at the bottom of the sheet that you are holding. This ensures that your idea is not lost or forgotten. The facilitator of the brainwriting session now collects the lists and congratulates teammates for coming up with dozens of insightful ideas in just a few minutes. The brainwriting session is now officially over.

Finer Points of Brainwriting

Brainwriting works well because it incorporates the best features of several different idea-generation techniques. First, brainwriters are able to work at their own pace, in the privacy of their own minds, without outside interference. When ideas are slow in coming, however, brainwriters can tap into outside stimuli. Another valuable feature of brainwriting is that everyone participates equally in a noncompetitive environment. Power or status differences within the group are neutralized effortlessly, as are conflicting personality traits of individuals. Teammates who are quiet, loud, shy, or overbearing are all equally able to make meaningful contributions.

Because brainwriting relies on anonymous inputting of ideas, generated ideas become the collective property of the group rather than of the individuals who offered them. This collective ownership of ideas erodes the power of cliques, voting blocks, and power players who may dominate a group. Collective ownership also promotes an environment where participants are free to express new ideas without fear of appearing stupid. This is healthy for productivity and interpersonal relations.

Brainwriting (as with all other idea-generating techniques) works best when participants let their imagination run wild. So be playful! Be outrageous! Have fun! Splash your ideas on paper without editing—anything and everything is fair game! And don't waste energy worrying about gram-

mar, spelling, or complete sentences—write down only enough to capture each thought so that you can quickly move on to your next thought. To promote a "what-the-hell" spirit even further, remember that *everyone* in brainwriting is supposed to let down their defenses and stop worrying about what others might think—those who *don't* let their hair down are the oddballs! Remember also that wild ideas are as good as safe ideas, and that raw, unpolished ideas have as much value as finely tuned ideas. Last, remember that the sole purpose of the brainwriting technique is to generate lots of ideas. At this point, quality of ideas is not relevant—sorting the wheat from the chaff comes later.

Suggestions for Successful Brainwriting

- Convert any objective into a brainwriting challenge by placing "*In what ways might we . . .*" in front of the objective.
- Emphasize playfulness and fun; brainwriting works best that way.
- Remember that wild ideas unleash entirely new ways of looking at things—so don't hold back!
- Resist the temptation to judge the goodness, badness, or craziness of ideas when brainwriting. Remember that brainwriting is a *creative* thought process that is hampered by judgment.
- *Quantity* of ideas, not quality, is the goal of brainwriting.
- When you start to bog down so that new thoughts are not coming quickly, switch lists with someone immediately; don't dilly dally. If the newly received list does not stimulate new ideas, switch again.
- Do not be timid when you request a switch. Sitting idle without new ideas is a waste of time.
- Be responsive when someone requests a switch; quickly finish whatever you are writing and then switch immediately with that person.
- When someone requests a switch, it is not necessary for everyone in the group to switch. Switch when you are ready, or when an individual asks you to switch. When in doubt, switch sooner rather than later.
- Keep each brainwriting session short, no longer than ten minutes. Tell participants at the start of each brainwriting session that the exercise will last only ten minutes. Stick to that time frame.
- When sharing ideas at the end of a brainwriting session, share ideas quickly. If the entire sharing process takes longer than a couple of minutes, you are moving too slowly.

Brainstorming

All people trying to solve problems know about brainstorming, or think they do. That is precisely why brainstorming is so often a waste of time. As originally conceived and as currently practiced by trained problem solvers, "brainstorming" is an effective, free-thinking group process that is carefully guided by a few strict rules. When the rules are not observed, however, "brainstorming" typically degenerates into a showcase of ideas from the group's dominant members.

To brainstorm effectively, participants must buy into four basic, inviolable precepts:

- Critical attitudes, judgment, and evaluation of ideas are absolutely off limits.
- Freewheeling is key—anything and everything is fair game. All ideas must be viewed as good ideas, no matter how zany or half-baked they may seem.
- Quantity breeds quality—the more ideas the better. Good ideas come from lots of ideas.
- Piggy-backing on ideas is good. New, better ideas often come from combinations, elaborations, or improvements of earlier ideas.

How to Make Brainstorming Work or Fail

Brainstorming can be exciting and productive, but only if personalities and differences in rank or authority of the brainstormers are equalized. Effective brainstorming results from good leadership and the right mix of players. If either variable is lacking, abandon thoughts of brainstorming and try brainwriting or the improved "nominal group technique" (p. 117) instead. A poorly conducted brainstorming session (the norm rather than the exception) is exasperating to all but those who dominate. It's also remarkably ineffective at generating good ideas—a few people left alone to their own devices will usually do a better job.

If at all possible, be selective about who is in your brainstorming group and who is leading the brainstorming session. If you have no choice in the matter, it's probably best to forego brainstorming altogether and use a different idea generating technique instead. If you are able to select individuals for your brainstorming team, however, choose people who are not afraid to think in new ways, and who are somewhat knowledgeable about the problem area. Look for people who are playful and creative and who function well in groups. Seek participants with similar status/power levels so that no one feels subordinate.

People are most creative and freewheeling when their working environment is supportive and fun—where playfulness and off-the-wall thinking are encouraged. People are least creative when surrounded by wet-blanket personalities who dampen the energy level. For those reasons, avoid know-it-alls, naysayers, and others who are quick to judge. And don't feel guilty about excluding them from your brainstorming sessions: Their time to make valuable contributions will come later.

How do you avoid hurt feelings when some people are left out? And how do you access ideas of those who are not part of your selected brainstorming group, but who probably have valuable insights? The answer is to use different idea-generating strategies for different groups of people. Use your hand-selected brainstorming group to jump-start the idea-generation process, and then circulate the grand list of ideas generated by the brainstorming group to interested outside parties. Ask those not in the brainstorming group to add additional thoughts and ideas to the brainstorming grand list, and to return the augmented grand lists to you by a specified time.

Proceeding in this way has several benefits: it provides difficult personalities with an idea-generation focus; this minimizes tangential ranting and raving. It also protects brainstormers from wet-blanket personalities, so that your most imaginative minds are not on the defensive. Third, it allows everyone to participate at a personally acceptable comfort level. Last and perhaps most important, everyone contributes and feels part of the process. When it comes to implementing ideas and strategies later on, this is worth a great deal.

Another approach to effectively accessing ideas from a wide mix of people is to divide possible participants into different brainstorming groups, with similar personality types grouped together. Put your preferred brainstormers in one group, wet-blanket types in another group, aggressives in another, and timid folks in yet another. At best, grouping people by personality type brings out the creative best in everyone; at worst, it permits everyone to participate equally while protecting the environment of your preferred brainstorming team.

Planning a Brainstorming Session

To conduct a brainstorming session, assemble a group of four to seven free-thinkers and ask each brainstormer to read this brainstorming section. A shared and equal understanding of the brainstorming process makes sessions more successful. It also makes the job of the facilitator much easier.

Now find a good brainstorming facilitator. To be effective, a facilitator must understand and enforce the ground rules of brainstorming. A good facilitator also keeps things moving, ensures that no one dominates, and promotes a relaxed, supportive environment where ideas can flourish. A model facilitator is relaxed, friendly, upbeat, and supportive, but not afraid to step in as necessary to maintain a favorable brainstorming environment. Ideally, the facilitator also has some understanding of the challenge being addressed.[2]

A good recording secretary is the other key player in brainstorming. An effective recording secretary is able to grasp brainstormers' ideas quickly and to capture them on paper in few words, in handwriting that others can decipher. Since it's not the role of the recording secretary or facilitator to generate ideas, it's best to recruit a secretary and facilitator from outside the brainstorming group. If your brainstorming group is small (5–7), one person can serve as both facilitator and recording secretary.

Arrange seating so that brainstormers can easily see and hear one another. If your group is small and members are comfortable with one another, arrange yourselves in a circle. If the group is larger, or if people are less familiar with one another, try a semicircular seating arrangement with a flip chart in front. However you choose to arrange yourselves, remember that people feel more comfortable and less exposed when they are seated around a table (as opposed to sitting in the open). Tablecloths that extend over the sides of tables are appreciated by women who are wearing shorts, skirts, or dresses.

The facilitator initiates a brainstorming session by displaying the brainstorming challenge where it's highly visible to everyone. The challenge is generally written in the form of a brainwriting challenge (e.g., *In What Ways Might We . . . ?*). Prominently displaying the challenge helps brainstormers stay focused. Before inviting ideas, the facilitator briefly reviews the four

2. Bosses and others with dominant or aggressive personalities tend to be poor brainstorming facilitators (they're rarely able to subordinate their own outflow of ideas to the good of the group).

brainstorming precepts presented earlier, impressing upon brainstormers that quantity of ideas is what they're after. When sessions start slowly, facilitators can fuel the idea-generating process by introducing ideas of their own.

If the facilitator maintains proper decorum, brainstormers rarely have difficulty coming up with lots of ideas. The difficulty comes with managing the incoming ideas. How to do this? When someone says something that triggers a new idea within you, what should you do? You know that interrupting someone else is rude, but waiting until the speaker decides to stop talking means that the context, excitement, and value of your own idea may be lost. It's also likely that someone else will jump in ahead of you, further separating your idea from the context that made it exciting.

What to do? There is no universally appropriate way to handle this situation, but a good starting point is to recognize that brainstormers generate two different types of ideas, "hitchhikers" and "outliers." "Hitchhikers," also called "piggy backers," are ideas that are spin-offs from a recently presented idea. If you have one of these, try snapping your fingers so that the facilitator, or the person speaking, can get to you right away before the context for your idea is lost. "Outliers," the other type of ideas, are those that head in a direction that's quite different from what presently is being discussed. Outlier ideas usually shift the direction of idea flow, so it makes sense to put these ideas on hold until "hitchhiker" ideas are infused into the current idea flow. Once hitchhiker ideas have been introduced, outliers then become fair game. (To ensure that your outlier ideas are not forgotten while you're waiting to introduce them, record them on paper. This preserves your outliers while simultaneously giving you mental freedom to rejoin the brainstorming process).

Facilitators of small brainstorming groups (4–6 members) sometimes forego formalities altogether and keep brainstorming an open dialogue, interruptions and all. This maximizes spontaneity and excitement in the group, but the facilitator must watch for overbearing personalities and long-winded oratories. Remember that it's always easier to set strict rules at the beginning and relax them later on than the other way around.

After fifteen minutes of brainstorming, when there is a lull in idea input, the facilitator should request a "round-robin," where every brainstormer—dominant and submissive—is given twenty to forty seconds of uninterrupted time to summarize the ideas that s/he considers especially fruitful. Only clarifying questions may be asked of the speaker during a round-robin, and these should be held to a minimum. With the first round-robin completed, brainstormers can now get back to work generating ideas. Restarting the brainstorming process is usually easy because of the idea stimuli introduced during the round-robin.

As the appointed end time approaches, or as brainstormers slow down noticeably (whichever comes first), the facilitator should request another round-robin to conclude the brainstorming session. As before, brainstormers use their uninterrupted time allocation (twenty to forty seconds each) to summarize ideas that they find worthy of further consideration. This last round-robin provides nice closure to the session.

Some Brainstorming Tips

Remember to break complex problems into smaller, more manageable challenges before trying to brainstorm solutions. Better yet, use your S-M-A-R-T objectives as your brainstorming challenges, taking the objectives one a time. Here are some other tips:

- DO remember that the purpose of brainstorming is to generate ideas, the more ideas the better.
- DO encourage off-the-wall ideas; they fertilize others' imaginations.
- DO brainstorm around a clearly stated and understood question, challenge, or objective.
- DO set a start and end time for each brainstorming session and hold strictly to it. Forty minutes is usually more than enough time if the session is conducted properly.
- DO select someone to facilitate/record ideas each time your group brainstorms. Ideas that are not recorded are quickly lost and forgotten.
- DO record all ideas on a flip chart where everyone can see them, and where misrecordings can be identified and corrected.
- DO create a friendly, upbeat working environment so that brainstormers feel welcome, supported, and important.
- DO ensure that everyone has a pressure-free opportunity to contribute.
- DO include a couple of round-robins in each brainstorming session.
- DO remind participants of the DOs and DON'Ts of brainstorming before each brainstorming session.
- DO tape-record round-robin summaries for future reference. The sheer volume of thoughtful ideas may overwhelm the recording secretary.
- DON'T permit individual(s) to dominate.

- DON'T permit long-winded soliloquies (a contributor should never need more than 30 seconds to input an idea).
- DON'T evaluate ideas when brainstorming; save that for later.
- DON'T brainstorm in large groups. Five or six brainstormers works well; more than twelve is usually an exercise in frustration. Break large groups into subset groups, making sure that each subset group has its own facilitator and recorder.
- DON'T hold long or open-ended brainstorming sessions; always conclude a session before brain death sets in. Insufficient time is usually better than too much time, for it motivates brainstormers to engage seriously and immediately.

Positive/Negative Forces Analysis (= Force Field Analysis)

Positive/negative forces analysis helps you break problems into smaller, more manageable pieces so that you can generate strategies for the resulting pieces. The technique is most usefully applied to problems that are troubling or bothersome.

Troubling problems have both good and bad forces acting on them. Negative (bad) forces fuel the troublesome nature of the problem and make it worse; positive (good) forces protect the problem from becoming more troublesome than it already is. The positive/negative forces analysis technique works by identifying and weakening the negative forces (so they are less bad), and by identifying and strengthening the positive forces (so they are more good).

To illustrate the use of this technique, let's say that your problem is, "Our organization's membership is declining" (see table 7.1). Begin by displaying this problem prominently so that all members of your problem-solving team stay focused on the problem at hand. Now dedicate one flipchart to "negative forces" and another to "positive forces."

First brainstorm or brainwrite negative forces (forces or factors that may contribute to membership decline), and record them on the "negative forces" flipchart. Then brainstorm or brainwrite positive forces (forces or factors that keep membership from declining more than it is). Record these on the "positive forces" flipchart. If a negative force is identified while brainstorming positive forces, or vice versa, simply record the force where it belongs.

Table 7.1.

Positive/Negative Forces Analysis

Example of how a positive/negative forces analysis (also called "force field analysis") is used to break troublesome problems into more manageable pieces for the purpose of stimulating new ideas and strategies. Remember to define the problem clearly, as a complete sentence. *Positive forces* are those that act to sustain or further the desired goal; *negative forces* do the opposite.

The Problem: Membership in our organization is in decline.

Positive Forces	*Strategies to Strengthen These Positive Forces*
• the organization has an important mission	1. issue press releases on issues so the organization gets some press
	2. have presence (e.g., info booth, pamphlets) at important events
	3. sponsor or cosponsor well-attended, cause-related events
• the newsletter is widely read and quoted	1. use newsletter to recruit and retain members
	2. increase circulation of newsletter (e.g., send copy to relevant organizations)
• members receive discounts at bookstore	1. create additional membership benefits
	2. advertise/promote membership benefits better
	3. create special trips and workshops for members
	4. make special organization tee-shirts
	5. have an organization logo contest

Negative Forces	*Strategies to Weaken These Negative Forces*
• no one is in charge of membership	1. appoint someone to be in charge of membership
	2. recruit a volunteer to be in charge of membership
	3. make everyone responsible for membership
• there's no volunteer help for membership recruiting	1. put someone in charge of volunteers
	2. use the Internet to recruit volunteers
	3. use newsletter to rally the troops
• the organization is perceived by many as out of touch with reality	1. align the organization with a respected, mainstream group
	2. take a very reasonable, high profile stand on an issue (and make sure the position gets press)
	3. use newsletter to trace how the organization has evolved from a radical fringe group to where it is today

Tape and display completed flipchart sheets of positive forces on one wall; display sheets of negative forces on a different wall. You now have captured the factors that collectively make the situation the way it is.

The displayed positive and negative forces identify situations that you would like to change, by making them either stronger (positive forces) or weaker (negative forces). Not all of the forces have equal impact on the status quo, of course, so first focus on those forces that have the greatest positive and negative impacts. Translate each of these especially important forces into a desired outcome (a S-M-A-R-T objective). To do so, decide how much the negative or positive force needs to change for you to feel that you've succeeded in dealing with that force.

Your specific, measurable, attainable, reasonable, and timed objectives provide targets for how you want things to change. With these desired outcomes clearly articulated, you can move on to strategies: "How might I move this force, from where it currently is, to where I want it to be?" You can generate strategies to strengthen each positive force, and to weaken each negative force through brainstorming, brainwriting, or through other techniques described in this chapter. After generating a bunch of strategies for each positive and negative force, you then can act on strategies that seem especially promising. Techniques for evaluating and selecting those that are most promising are described in the next chapter.

Developing a plan for implementing strategies is often easier said than done. If you don't know where to go from here (i.e., you want to implement the strategy but don't know how), you need to tap into your creative energies once again. By now you may recognize that this is easy: simply precede each elusive strategy with "In what ways might we . . . ?" For example, "In what ways might we use our newsletter to recruit and retain members?" Now brainwrite, brainstorm, or use another idea generation technique in this chapter to help you develop plans for implementation. If this all seems rather involved, it's not (see pp. 147–148 for an example of this stepwise idea generation process).

Desired Features

"Desired features" is an idea-generation technique that works by fine-tuning your objectives. To illustrate, let's say that your organization's newsletter isn't as good as you'd like it to be. (This is a problem because your desired outcome—a great newsletter—is different from the current situa-tion—a mediocre newsletter.) Let's also say that, after some effort, you have set the following S-M-A-R-T objectives:

- To have in hand a printed, complete, typo-free draft of the next issue of the newsletter by October 15.
- To include at least two new, regular features in the next issue.

These objectives are helpful because they identify what you want in the end. They do not ensure a better, more effective newsletter, however, be-cause they do not spell out the specific changes that need to occur to move the newsletter from "mediocre" to "great." Recognizing these limitations, you need to identify additional S-M-A-R-T objectives that articulate the qualities you want in your next newsletter. You then can figure out ways (strategies) to make them happen. The desired features technique effec-tively identifies desired features and strategies that might be used to attain them. Use this technique whenever it's unclear that achieving your stated objectives will in fact lead you to a successful outcome.

To employ the desired features technique, begin by making a list of all the great characteristics that you'd like your newsletter to have (table 7.2). Now list ideas that might advance each of these desired features. Remember, as with all idea-generation techniques, that the more ideas the better.

Sometimes it's easy to identify a desired feature (what you want in the end) but difficult to know how to get there. When you're in this situation, tap your creative powers by generating strategies through brainwriting or brainstorming. Simply preface each desired feature with, "In what ways might we . . . ?" (e.g., "In what ways might we make the newsletter attrac-tive?" "In what ways might we make the newsletter inexpensive?"). Gener-ate strategies, one desired feature at a time, to ensure that the strategies tar-get a specific desired outcome rather than the amorphous problem of a mediocre newsletter. Select strategies for desired features that are easy and inexpensive to implement.

Using the desired features technique is not how most people would ap-proach the problem of a mediocre newsletter. Most people would instead look to other newsletters for inspiration, the thinking being that there is much to learn from what others have done. This certainly is true, but there are disadvantages to conditioning your mind prematurely to a certain way of thinking. If you are looking for truly new, exciting ideas, the freshness and individuality of your own imagination will take you in directions prob-

Table 7.2.
Desired Features Technique

Example of how the desired features technique might be used to generate new ideas and strategies. As always, the problem definition is a clear, declarative statement of the situation that you would like to be different.

The Problem: Our newsletter is not as good as we'd like it to be.

We Would Like It to Be		
Substantive and credible	*Engaging*	*Attractive*
Strategies for How Each of the Above Desired Features Might Be Achieved		
• include pieces by respected dignitaries • cover all sides of issues • include expert reviews • include bibliographies	• get school kids to color pictures; include pictures in newsletter • profile unusual people, events • feature a professional writer • include how-to-do techniques	• hire/recruit designer/artist • include color photos • use desktop publishing • use high quality paper • hand-color pictures

ably quite different from those manifest in other newsletters. So, before contaminating your imagination with how others have done it, first mentally fashion your conception of desired features and how to attain them. With these thoughts recorded, then review other newsletters for nifty ideas that you haven't thought of. Borrow ideas from other newsletters and incorporate them into your newsletter conception to create the best newsletter ever: one that combines your best (new) ideas with the desirable features of other newsletters.

Mind Mapping

Mapping your mind through sketches, flow charts, and balloon diagrams stimulates creativity and leads to new thought processes. So, always try to make a picture of the problem and how you are approaching it. You'll be pleasantly surprised at how much you learn from the process and how many new ideas spring to mind.

You're mistaken if you think that sketching ideas isn't practical for esoteric, theoretical, or abstract problems. Trying to represent these problems visually forces you to package the problem in unconventional ways. If you're looking for new ideas or new ways of approaching the problem, that's just what you want.

Flowcharting

One common and somewhat formalized mind mapping technique is the flowchart. By organizing pieces of your problem and approach into a sequential flow, you can examine the pieces and the connections and speculate on alternative pieces and connections. After you've identified pieces of the problem and your approach, and have ordered them into a flowchart that captures your version of reality, ask yourself if all of the pieces are necessary. Then consider what might happen if certain pieces were left out. What would happen if the order of events were changed? How could connections between events be changed? Asking these questions generally prompts you to change the pieces or the order of the pieces in your flow chart. This invariably leads to new ideas.

Here are a couple of examples of how flowcharting might be used to stimulate new approaches to the problems.

Problem 1: Our representative is not honoring her campaign promises.

Possible Objective for Problem 1: Have at least two of her promises honored by August 1.

Flowchart 1: If you were to make a flow chart of the problem and your possible courses of action, it might look something like this:
   (1) Representative makes campaign promises.
      ↓
   (2) We support her with volunteer help and $$.
      ↓
   (3) She's elected.
      ↓
   (4) We rejoice.
      ↓
   (5) We wait for desired actions.
      ↓

(6) Desired actions don't happen.
↓
(7) We send letters expressing our disappointment.
↓
(8) We receive noncommittal response.
↓
(9) Desired actions still don't happen.
↓
(10) We're frustrated and discouraged.
↓
(11) We search for a way to force action.

The flowchart above may accurately capture what has happened and how you've proceeded, and your initial scan of the flowchart might reassure you that your actions have been reasonable. But your reasonable actions haven't worked, so you need to try something new. To generate new strategies for tackling the problem of a do-nothing representative, first look at each flowchart piece, one at a time. Could the piece be changed or manipulated in some way, even if the modification doesn't make immediate sense? (The answer should always be "yes.")

Here are some new ideas and strategies that might be generated by reconfiguring the first two events in the flowchart above:

1. "Representative makes campaign promises."

- Our organization makes campaign promises in the future (e.g., we publicly state ramifications if a candidate we support does not follow through).
- Our organization publicizes campaign promises that have not been met.
- We threaten to launch a media blitz if media promises are not met.

2. "We support her with volunteer help and $$."

- We help her implement her campaign promises.
- We won't support her next campaign.
- We threaten not to support her next campaign.
- We send a letter to her campaign donors revealing her failure to honor campaign promises.

Here's another example of how a flow chart could be used to generate new strategies for attacking a problem:

Problem 2: I don't know how I'll complete my field sampling and data analysis before snowfall.

One possible objective for problem 2:
Complete all inventories of tree seedlings by October 1.

Flowchart 2: If you were to make a flowchart of your desired outcome and the strategies you're thinking of using to get there, your flowchart might look something like this:

(1) Get equipment.
↓

(2) Learn how to identify the different tree seedlings.
↓

(3) Load gear in car each field day.
↓

(4) Drive to field site.
↓

(5) Carry gear to site.
↓

(6) Select plots using random numbers.
↓

(7) Install marking posts for each plot.
↓

(8) Identify and count tree seedlings and record numbers on clipboard data sheets.
↓

(9) Enter data into computer.
↓

(10) Analyze data.

Here are some new ideas for tackling this problem that might be generated by reconfiguring events in the flow chart above:

• Enter data directly into a computer in the field (forego the time-consuming step of recording and transcribing data from data sheets).

• Don't wait to start selecting sites and installing posts; do this right away while you're learning to identify seedlings.

• Store gear at the site to save time.

• Camp out at each site to reduce travel time.

• Get others to help you, with each helper assigned a specific task that matches his/her skills.

Reconfiguring pieces of flowcharts can yield many new ideas, but it's sometimes difficult to imagine how reordering entries in a flowchart (or omitting entries altogether) could yield anything new. If this describes your experience, try using the "storyboard technique" to reorder (or eliminate) entries in your flowchart.

To use the storyboard technique, begin by transferring each flow chart entry to an index card—one entry per card. Now display the completed cards on a flat surface, as if you were playing solitaire; the resulting card sequence should represent your perception of the chronology of the situation.

To trigger new ideas, try changing the storyboard flow by changing the order of cards in the storyboard; also try removing cards. Do the changes alter how you think of the situation? Does the reordering of cards point you in new directions? Does removing one or more cards altogether change your perspective and prompt new ideas? Now rearrange the storyboard cards in different orders. Remember to try removing different cards from the sequence to see what ideas might emerge when steps are omitted.

Balloon Diagrams

The "balloon diagram" is another variation of mind mapping that also is effective at stimulating new ideas and strategies. It closely resembles "clustering," a technique used by writers to generate ideas and to organize thoughts (p. 69).

Balloon diagrams help you focus on manageable, bite-size parts of the problem that can be tackled one at a time (the divide and conquer approach). This is especially useful early on in a problem-solving effort because the diagramming process quickly exposes lots of thoughts and ideas. From this collection of mental meanderings you then can identify the ideas and problem components that are most promising.

Balloon diagrams are also effective in later stages of problem solving because they help you see connections, interfaces, and threads that tie together the different problem pieces. These insights generally yield new ideas and strategies to achieve your desired outcomes.

Here's how a balloon diagram might be used to generate new strategies for dealing with the following problem:

Our projected expenses exceed our projected income. Examples of
S-M-A-R-T objectives for this problem might be:
• Reduce expenditures by 20 percent starting next week.
• Increase income by 20 percent starting next month.
• Balance our budget by July 201X. (etc.)

The balloon diagram shown in figure 7.1 is a first iteration of how
you might think about the first objective, "reduce expenditures." That's a
helpful beginning to solving your financial problem because the problem
has now been broken into progressively smaller units—first into objectives
(such as the three above), and then, using the balloon diagram, into expenditure items. Each of the expenditure items in the balloon diagram can
easily be converted into a more specific objective (e.g., "reduce long-distance
telephone expenditures by 15 percent," "reduce postage expenditures by 25
percent"), and the more specific objectives can then be tackled individually
by generating strategies to achieve them through brainwriting, brainstorming, or other idea-generation techniques.

Balloon diagrams stimulate new ideas and strategies in other ways.
Some new ideas usually surface as you're creating the balloon diagram, others come from inspection of the completed diagram. For example, as you
jotted down "equipment" on this particular balloon diagram, it may have
occurred to you that perhaps you could reduce expenditures by borrowing
rather than buying equipment, or by sharing equipment with other users.
When new strategies such as these come to mind, record them below the
balloon diagram so that they're not lost and forgotten. Be sure to record all
ideas, even those that seem unworkable. Remember: all ideas have equal
value during the idea-generation stage. The goodness or badness of ideas is
evaluated later.

After completing a balloon diagram, you may spot some interfaces that
warrant further exploration. Perhaps the custodian or secretary, for example, could be trained to maintain specific pieces of equipment. That
would save the expense of paying outside experts. Or perhaps you could reduce telephone and postage expenditures by making better use of the In-
ternet. Or perhaps you could seek out volunteers who have skills that you
currently have to pay for in permanent staff or outside consultants.

It certainly may be that, in the end, you choose to implement none of
your newly generated strategies. That's okay. The idea generation exercise
was valuable nonetheless, for it opened your mind to new possibilities.

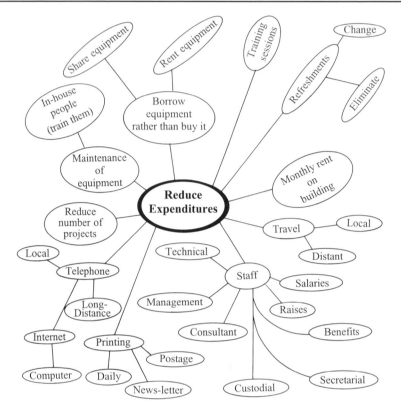

Fig. 7.1. Balloon Diagram of a S-M-A-R-T Objective
The Objective: To reduce expenditures by 20 percent starting next week.

Generating Ideas When Your Group Is Dysfunctional

The idea-generating techniques above work well if interpersonal dynamics within your group are healthy and members respect one another. With environmental problems, that's oftentimes more wishful thinking than reality. Group problem solving is often sabotaged by power struggles, status differences, cliques, and past interactions. If this describes your group, you need to employ an idea-generation technique that protects participants from one another so that generation of new, different ideas is possible. Two techniques are especially effective at this: "Brainwriting" (described earlier) and the "Nominal Group Technique" (described below).

The Nominal Group Technique

The nominal group technique (NGT) works well when members of the group have different status or power, or when they stand on very different sides of the problem. NGT also works well when some members of the group are hesitant to speak openly and freely, or are likely to "take sides." NGT neutralizes insecurities, predispositions, and differences in power by ensuring that people offer ideas under a cloak of anonymity.

NGT is not complicated or involved, but it often seems odd and unnecessarily regimented to those who are not familiar with it. Getting buy-in from strangers who are meeting for the first time therefore requires a convincing explanation of the process. Having each member of the group read this section of the book is a good way to bring everyone up to speed.

For NGT to be successful, a committed, disciplined facilitator is needed. This person need not be drawn from the group, but the integrity and fairness of this person should be beyond reproach.

How the Nominal Group Technique (NGT) Works

The facilitator jump-starts the NGT process by asking every participant to read the NGT section of this text before meeting. This preparatory step is critical: participants will be disinclined to buy into NGT if they don't understand how and why it works.

As another preparatory step, the facilitator should present in advance—if at all possible—the specific challenge for which strategies are needed (e.g., "In what ways might we resolve the conflict between ATV users and wilderness advocates?"). The facilitator asks participants to offer anonymous ideas before the meeting by providing each participant with a few standard index cards, and asking each participant to record on the card a few ideas triggered by the stated challenge. Participants deposit their anonymous ideas (written on index cards) into a "ballot box" that only the facilitator can access.[3]

On receiving ideas from group members, the facilitator then records the submitted ideas on sheets of flipchart paper (in private) so that the ideas can be presented during the group meeting. The facilitator also mixes in

---

3. Group members may submit typed or computer-generated comments to the ballot box, or send them electronically, so long as everyone provides feedback in an anonymous way.

ideas of his/her own to the flipchart sheets so that members will have a number of stimuli when they meet as a group.[4] Each idea is assigned a unique reference number.

At the appointed meeting time, members sit in a semicircle facing the facilitator, and a flipchart sheet is displayed which states the challenge for which ideas are needed. The facilitator then hands each participant several blank index cards and briefly reviews how the nominal group technique works. As with all idea-generating techniques, it's important that members are reminded to forestall all evaluation of the quality or practicality of ideas. Generating a large number of new ideas should be the sole concern and focus.

The NGT facilitator now asks each participant to suggest ideas for how the flipchart challenge might be met. To ensure anonymity of ideas, participants are asked to record their ideas on index cards, working privately and independently. After three to five minutes, the facilitator collects cards, face down, from participants. Every participant is reminded, just prior to the collection period, to submit at least one card to the facilitator. If a participant has no new ideas to add (and nothing written on an index card), that participant should submit an index card nonetheless. Ensuring that everyone submits a card—regardless of whether or not anything is written on it—protects the anonymity of those who do offer new ideas.[5]

After collecting the cards, shuffling them, and placing them aside temporarily, the facilitator now displays the flipchart sheets of numbered ideas that were scribed prior to the NGT meeting. Participants are asked to review the displayed flipchart ideas and to write down (on new index cards) any new ideas that come to mind.[6] While participants are working independently to record newly stimulated ideas on index cards, the facilitator and another scribe retreat to a private location,[7] shuffle the cards anew, and

---

4. Some facilitators forgo trying to get group members to submit cards before the meeting because it's rare to have many ideas submitted ahead of time. Requesting premeeting submission of ideas is still advised, however, because doing so enables the facilitator to offer ideas without anyone knowing where the ideas originated.

5. This procedure—that *every* participant submits a card, regardless of whether or not anything is written on the card—is essential to the NGT process, for it establishes a safe environment where participants dare to be honest and forthright.

6. Since the purpose of this (and all) idea-generating techniques is to generate as many different ideas as possible, participants shouldn't waste time repeating ideas that they already have offered during an earlier card session.

7. Scribes are encouraged to conduct their transcriptions out of sight of the rest of the group so that they are neither distracted nor distracting.

transfer ideas on the cards to new flipchart sheets.[8] To the extent possible, the scribes try to capture ideas in the author's language.

The transferring of ideas from cards to flipchart sheets usually takes several minutes, and it's likely that the outpouring of new ideas from participants will have waned by the time scribes have completed their transcriptions. A moribund output of ideas signals that it's time for the facilitator to collect cards from participants. As before, each participant submits a card (face down) to the facilitator, irrespective of whether or not anything is written on the card. As before, newly collected cards will be shuffled and scrutinized for new ideas. These new ideas are then assigned a unique number for future reference and are transferred to new flipchart sheets where they can be seen easily.

Transcription of ideas from the second set of cards often takes longer than you would expect, so it is wise to have several scribes (and flipcharts) on hand to speed along the transcription process. If participants find themselves sitting idle, waiting for the transcription to be completed, the facilitator should use the down time to make noncontroversial announcements. A short break also can be offered.

Once transcription has been completed, the facilitator asks participants to review items on the flipchart sheets for clarity. Any participant may propose a rewording of any item (regardless of who authored the idea) by suggesting a specific rewording. When a proposed rewording is offered, the facilitator asks participants if there is any objection to the proposed rewording. If there is no objection, the proposed rewording is made by lightly deleting the original (so that the original item is still visible) and replacing it with the newly worded version. If anyone objects to the suggested change, there is no further discussion or resolution: the original item is left unchanged and the proposed change is added to a flipchart sheet and treated as a new (numbered) idea.

The Challenging Part of NGT

With ideas prominently displayed on flipchart sheets, the time has come to discuss the ideas. The discussion phase can be hairy if there is discord in the group—which there probably is if you've opted to use NGT—so it's critical that the NGT protocol be followed scrupulously. This begins with

---

8. As with the first batch of ideas, each newly transferred idea is assigned its own unique number. Repeats of ideas are not given new numbers and they are not transferred to flipchart sheets.

the facilitator, who facilitates and manages discussion while at the same time completely withholding his/her own personal preferences and judgments. NGT participants must never wonder if the facilitator is favoring certain ideas over others or the NGT process will implode.

Before entering the discussion phase of NGT, the facilitator should spend a few moments reviewing the protocol and rules for this part of the process. Participants need to be reminded that:

The discussion phase of NGT is not a debate or lobbying session, and it is not a time to evaluate ideas, or to vote on which ideas should be retained or discarded. The discussion phase *is* for efficiently and respectfully exploring the merits and liabilities of each item on the sheets.

The facilitator reminds participants that grandstanding, speech-making and long-windedness are not permitted. Participants are also reminded that, once a point has been made, it should not be made again. Repetition is out of order.

All items on the flipcharts are open for discussion, so the facilitator must ensure that participants stay focused on the specific task at hand and that everyone plays by the rules. When someone strays from the appropriate protocol, which invariably happens, the facilitator must quickly and politely remind the wandering speaker to stay on track. A good way to do this is to ask, "Does the speaker have something new and specific to add?" The facilitator also should remind everyone that official evaluation of ideas comes later, after all items have been discussed.

Although all items may be discussed, there is no obligation to discuss an item if no one is interested in doing so. If the group is dealing with a large number of items, it's oftentimes wise to put a time limit for discussion on each item. This discourages long-windedness and repetition because members recognize that they must be efficient to make their points.

New ideas can be added to the flipchart sheets at any time, and the facilitator can ask for other rounds of anonymous inputting of ideas whenever that seems useful. Often, there are so many ideas on the flipchart sheets that there will not be time to discuss items fully during the planned meeting time. As soon as this reality emerges, the facilitator should direct the group to make a binding decision on how to proceed: Should we schedule another meeting? Should we extend this meeting? Should we limit discussion of each item to a specified period of time to ensure that all items are discussed equally? The group should agree on a strategy to avoid being strapped for time at the end of the meeting. When time is short, the tendency is to blitz through remaining items for the sake of completion. Resist that temptation.

When the discussion phase has been completed, thank participants for helping put forth lots of new ideas, and for helping everyone see the situation from different perspectives. The stage is now set for making good decisions about which ideas to keep, and which to discard (the focus of chapter 8).

When You're Really Stuck—Brain Fertilizers

The biggest challenge faced by problem solvers working alone is invoking outside stimuli to awaken new thought processes. People working together also can become stuck on tired ideas and strategies, however. Brain fertilizers, such as those below, are godsends when you find yourself in this situation.

Incubation

An especially effective but elusive idea-generation technique is "incubation." Incubation is elusive because you can't force it, you never know what (if anything) will come from it, and you don't know when or where it might produce something useful. You can only create an environment for incubation to flourish and see what happens.

Incubation, also called "fermentation," is a key ingredient in environmental problem solving because it provides time for your thoughts and ideas to mature. As a testament to the value of incubation, recall that you sometimes experience breakthroughs in your thinking when you least expect it, or that you magically remember a forgotten name long after you've moved on to something else. Incubation works because your subconscious mind labors over problems, even when the rest of you isn't.

Table 7.3 offers a little quiz that should convince you to take incubation seriously. How many answers came to mind immediately? How many answers miraculously appeared later, completely out of the blue? Most people

Table 7.3.

How Your Mind Works When the Rest of You Isn't

Take the following test, recording answers so you know what you've answered and what you haven't. Each set of initials indicates a well-known expression.

The Test: Identify the missing expressions.

Example: 16 = O. in a P. (*Answer:* Ounces in a Pound)

1. 26 = L. of the A.	13. 88 = K. on a P.
2. 7 = W. of the A. W.	14. 20 = F. and T.
3. 1001 = A. N.	15. 4 = Q. in a G.
4. 12 = S. of the Z.	16. 24 = H. in a D.
5. 54 = C. in a D. (with J's)	17. 1 = H. on a U.
6. 9 = P. in the S. S.	18. 5 = D. in a Z. C.
7. 50= S. on the A. F.	19. 57 = H. V.
8. 13 = a B. D.	20. 7 = T. and E. P.
9. 32 = D. F. at which W. F.	21. 1000 = W. that a P. is W.
10. 2001 = a S. O.	22. 29 = S. in F. in a L. Y.
11. 90 = D. in a R. A.	23. 40 = D. and N. of the G. F.
12. 200 = D. for P. G. in M.	24. 6 = S. on a C.

find that certain scenarios (e.g., taking a walk, taking a bath, running) seem favorable to incubation breakthroughs. Keep track of when and where mental breakthroughs occur. You then can recreate this same environment when you need the powers of incubation most.

When you've reached a dead end and are tired, discouraged, or frustrated, take time out to get away from the problem entirely and let incubation do its work. Do something relaxing, mildly physical, and completely unrelated to the problem. Spend some time alone. If you're too stressed about the problem to allow yourself to relax, make peace with yourself by officially taking time off to let incubation do its magic.

Here are some suggestions to make incubation work for you better:

- Mindless physical tasks seem to stimulate incubation. Do chores, go for a walk, do a workout. Sitting, watching TV, and reading don't work as well for most people.

- Plan ahead. Work backwards from deadlines and schedule a block of time (half a day, minimum) where you get away from the problem entirely and relax, by yourself. If you can, try to get away from the problem for more than one day. Thorny problems often need time.

- Keep track of when, where, and what you're doing when incubation works for you. Recreate that setting to optimize your incubation time.
- Be patient. You can't force incubation, you can only create an environment where it is likely to flourish.

Synonyms

One simple but effective way to jostle your mind into new ways of thinking is to reformulate your problem definition using "Synonyms." At the top of a new sheet of paper, write out the problem as you have defined it. Leave space between words so the problem definition is spread out across the page. Now consult your thesaurus for synonyms of each word in the problem definition. List these synonyms under the associated word in your problem statement to create a page of columns of synonyms (table 7.4). As you'll discover, the process of finding and considering synonyms will trigger new ways of thinking about the problem, and new ideas and strategies. Record these at the bottom of the sheet so that you don't lose them.

Forcing Relationships

Often, reformulating the problem definition with synonyms is sufficient to sprout generation of new ideas. You can take the synonym technique a big step further, however, by "forcing relationships" between synonyms.

To begin, first identify the two or three key words in your original problem definition that capture the essence of what the problem is really about; usually these words are one or two nouns and a verb. Now try different combinations of synonyms of the different key words to create new problem definitions. Do these new articulations of your problem stimulate new ideas and perspectives? If so, record them so that they're not lost.

Here are some (alternative) problem definitions that were generated by forcing relationships of synonyms from table 7.4:
- Too few river guides cultivate beneficial (environmental) conduct.
- Too few river guides cultivate appropriate (environmental) behavior.
- Too few river guides maintain appropriate (environmental) behavior.
- Too few river guides promote better (environmental) policy.
- Too few river guides restore desirable (environmental) management.

Table 7.4.
The Synonyms Technique

Example of how the synonyms technique stimulates new ways of thinking about a problem. Words at head of columns are key words for which synonyms were sought.

The Problem: Too few river guides. . .

Promote	*Good*	*Stewardship*
cultivate	beneficial	conduct
advance	appropriate	behavior
forward	innocuous	management
enhance	satisfactory	economy
foster	desirable	policy
improve	better	ministry
restore	virtuous	statesmanship
maintain	unobjectionable	oversight
correct	moral	guidance
(etc.)	(etc.)	(etc.)

As the examples above demonstrate, "forcing relationships" casts your problem in a number of different lights, all of which help you do a reality check on the correctness of your problem definition, your objectives, and your problem's constraints. For example, is the "river guide problem" really about guides doing a poor job of controlling their clients' behavior on the river? Or is the real issue that river guides should be taking an active role in promoting environmental awareness, restoration, or management (but aren't)? Or is the "river guide problem" really about guides being inconsiderate slobs, and allowing their clients to be inconsiderate slobs also?

As these forced relationships demonstrate, slight shifts in wording can lead to dramatically different takes on the problem, the goal, the objectives, and the strategies that you might employ to meet those objectives.

Assumption Reversal

As with the techniques above, the purpose of the "assumption reversal" technique is to break through mental gridlock. Assumption reversal is simple: Begin by listing *all* assumptions you are making about the problem, being sure to include those that you take for granted. When you feel your list is quite complete, scan notes in your "boundaries and limitations" folder for assumptions that you may have overlooked. Now reverse each assumption (any way you can) to see what ideas emerge from the reversal. For example, let's say that the problem is that: *RV users are harassing wildlife*.

Assumptions	Assumption Reversals
RV users are thoughtless slobs	RV users are thoughtful and respectful
Harassment harms wildlife	Harassment helps wildlife
Wildlife = birds and mammals	Wildlife = people, insects, plants, soil
RV users are mainly men	RV users are mainly women and children
Wildlife is bothered by RVs	Wildlife likes RVs
RV users need to stop	More RV should be encouraged to do what they do
RVs are loud and obnoxious	RVs are quiet and pleasing
RVs are polluting	RVs contribute to pristine resources

For brevity, only one reversal was shown for each assumption; in practice, you should list as many reversals as you can.

The assumption reversals above hopefully stimulate new perspectives and ideas. For example, the first reversal above might make you think that perhaps we could find and recruit RV users who *are* thoughtful and respectful, and convince them to police those RV users who are not thoughtful and respectful. The second reversal might help you remember that in some places wildlife harassment might be beneficial, such as where deer are devastating natural plant communities, or are threatening rare and endangered species. In these places, RV use could be encouraged and special RV use areas could be designated and promoted. This would acknowledge and support the rights of RV users, simultaneously making it easier to justify a compromise that keeps RV users from harassing wildlife in certain sensitive locations. The third reversal (wildlife = people, insects, plants, soil) might make you think that people's outrage over the intrusiveness of RVs might be harnessed to discourage RV use in some locations, thus protecting wildlife in those areas from harassment.

Restructure Attributes

When your idea-generating output is lower than a worm's belly in a ditch, you need to break from old ways of looking at the problem and try something new. That's what "attribute restructuring" does. Each problem and attendant solution is defined by a particular suite of characteristics, its attributes. Changing the problem by changing its attributes stimulates new ideas and reveals new possible solutions.

To illustrate, let's say that you want to solve the following:

Problem: If current trends continue, the current landfill soon will be unable to accommodate the community's waste.

Your Goal: Come up with a plan that will accommodate the community's future waste.

Your Objectives:
- Develop a plan by October 15 that will handle the community's waste stream for the next ten years.
- Develop a waste disposal strategy by the end of next year that holds costs to consumers at current levels.
- Develop a waste disposal strategy by the end of the year that residents find easy to implement.

With your problem, goal, and objectives on paper, you now turn your attention to identifying the problem's attributes.

Attributes of the Problem: Here are some of the attributes that we might have come up with:
- All residents of the community use the landfill.
- The main waste stream entering the landfill is domestic trash and garbage, old furniture, old tires, brush from land clearing, discarded building materials from contractors, and carpet remnants from the local carpet store.
- Jack's Trash Service dumps its trash in the landfill.
- Jim Smith watches over the landfill, buries trash periodically with soil to keep pests under control, and scavenges items of value.
- Jim is paid $20,000 a year for his services; he keeps whatever income he can generate from selling his scavenged items.

- The landfill is an abandoned gravel pit and it's surrounded by active gravel pits.
- The landfill is 1.6 acres in area.
- Anyone can dump trash in the landfill.

Examples of Ideas Stimulated by a Restructuring of the Problem Attributes

- Maybe not everyone should use the landfill.
- Perhaps more people should use the landfill—the town should join forces with neighboring communities and develop a regional waste plan, or a mega-landfill.
- Maybe the waste stream entering the landfill should change—maybe the town's landfill should "specialize" in certain types of waste, with neighboring towns specializing in different types of trash (work cooperatively).
- Maybe landfill users should be managed differently—for example, contractors and businesses should be charged a fee for disposing their wastes.
- Maybe contractors, businesses, and trash collecting services should be required to dump their wastes elsewhere.
- Perhaps the community should charge a fee for each bag of trash dumped into the landfill. This would encourage residents to recycle and reduce trash input.
- Maybe Jim should set up a recycling station at the dump, where every visitor has to contribute recyclable items before dumping other stuff into the landfill.
- Maybe Jim's position should be put out for bid: whoever comes in with the lowest bid may set up a recycling station at the dump and reap all of the profits (promoting competition and imagination).

You may have noticed that a restructuring of attributes sometimes drifts into the domain of constraints, limits, and boundaries (chapter 6). Be alert to this possibility so that your limits and boundaries always fit the rest of your problem. If they don't fit, change them so that they do.

Exaggerating Objectives

Exaggerating your objectives can open your mind to entirely new types of strategies. To illustrate how "exaggerating objectives" works, let's say that you are concerned about the rapid and haphazard development occurring in your town and you wish to do something about it. After clearly defining the perceived problem and goal, and clearly stating your S-M-A-R-T objectives, stretch or exaggerate each objective in all imaginable directions.[9] These new, often preposterous objectives provide cerebral fodder for new ideas. For example:

The Problem: Excessive new home construction is changing the character of our town.

The Goal: To preserve the character of the town.

Possible Objectives:

1. Institute land-use planning in the town by January 1, 201X.

2. Find alternative means of covering the property tax base, starting with the next fiscal year.

3. By the end of the year, articulate a clear, compelling argument for "average" town residents about why the current level of home construction is not desirable.

Exaggerated Objectives for Objective 1:
- Halt discussion of land-use planning.
- Become militant about the virtues of land-use planning.
- Make kids militant about the virtues of land-use planning.
- Encourage people to do anything they want, anywhere they want to do it.
- Convince people why land-use planning is undesirable.
- Institute land-use planning in the town IMMEDIATELY.

Exaggerated Objectives for Objective 2:
- Do away with taxes.
- Increase taxes.
- Give tax rebates.
- Reduce the town's tax base.
- Have new construction efforts pay for all of the town's taxes.

9. Don't worry about, or even try, making your exaggerated objectives S-M-A-R-T.

- Find alternative means of covering the property tax base, starting NEXT WEEK.

Exaggerated Objectives for Objective 3:

- Generate a clear, compelling argument as to why the current level of home construction *is* desirable.
- Generate a compelling argument for all people everywhere as to why the current level of home construction is *not* desirable.
- Be radically emotional and irrational about land-use planning.
- Articulate a clear, compelling argument by TOMORROW for "average" town residents why the current level of home construction is not desirable.

Here are some strategies for solving the problem that were stimulated by the exaggerated objectives:

- Show landowners the property value benefits of open space and tasteful, planned development.
- Write newspaper article warning residents about the increased need for town services (teachers, schools, police, road maintenance, fire, sewer, etc.) with increased population pressure.
- Discuss with school children the issues related to haphazard development (if you get kids on your side, you're halfway there).
- Get town leaders on your side so that you're not a lone voice in the wilderness.
- Work with developers rather than against them.
- Become a developer yourself (an "eco-friendly" developer).
- Clearly understand the concerns and views of those opposed to land-use planning so that each concern and view can be addressed.
- Run for office so that you have a political voice in land-use planning decisions.
- Seek help and support from like-minded groups and organizations outside your town.
- Find out how other towns have handled this problem.
- Get the media interested in your quest.
- Find out what has happened elsewhere when development was unchecked, and show townspeople what they're in for if haphazard development isn't curbed.
- Give tax breaks to landowners who don't sell out to developers.
- Do a cost-benefit analysis of the tax gain resulting from each additional

house and family (compare it against the increased town expenses resulting from population and housing growth).
- Remind townspeople that their traditional uses of the land for hunting, fishing, and other recreation will be lost if the land is locked up or fragmented by new construction.
- Hasten changes through a referendum.

Serendipity

Serendipity is the goddess of creativity but many environmental problem solvers are oblivious to its presence or power. You can become more receptive to its influence on a day-to-day basis by slowing down and paying more attention to what's around you. Adopt an attitude that there are great strategies out there to solve every problem, you only need to find them.

You needn't sit on the sidelines waiting for serendipity to strike, you can promote it with little effort. For example, if you're struggling to design or build a piece of equipment, or if you're needing to find an inexpensive alternative to a manufactured piece of equipment, try visiting a big hardware or craft store. Farm stores, junkyards, and landfills also work well. Roam the aisles, every aisle, taking your time. Pick up gizmos and look them over, fiddle with everyday building materials and let your imagination run wild. Make notes and sketches in a notebook so that your ideas aren't forgotten. You'll be amazed at what you come up with.

You can promote serendipitous discoveries for other types of problems by similarly seeking out possible depots of brain fertilizer. Try wandering through a large office supply and stationery store when you're struggling with organizational problems. Wander through giant department stores such as KMart and WalMart when you don't know where else to turn. Look through gear catalogs, especially those for "working people."

If you're completely lost or stuck, discuss your problem with a professional reference librarian. Reference librarians know, or know how to know, just about everything.

Disassemble/Reformulate the Problem

Thought processes, especially idea generation, are constrained by how you define the problem, the goal, the objectives, and the constraints. When you're really stuck, try rethinking your problem, goal, objectives, and constraints one more time using the "repeat why" technique (p. 64), the 5 Ws and H technique (p. 88), and the boundary examinations technique (p. 89).

Some Important Reminders Before Moving On

Whichever techniques you use to generate ideas, strategies, and possible solutions, keep in mind (1) that the purpose of idea generation is to create lots of ideas; (2) that zany ideas are as useful to the idea-generation process as sensible ones; and (3) that you need to treat idea generation (the "S" in DOC'S KEY) as an entirely different step in the problem-solving road map from idea evaluation (the "K" in DOC'S KEY).

Wise Words from Old Pros

- Most of the techniques described above can also be used to generate ideas when you're working alone. Discussing your problem with non-specialists is always helpful, and positive/negative forces analysis, desired features, and mind mapping consistently work well at getting ideas on paper. Other techniques that work well when you're working alone include freewriting and clustering. Try them all.
- Always include "do nothing" as one of your possible strategies (i.e., solutions).
- Remember that new ideas can be new connections or combinations of old ideas.

Also remember that:
- The best way to have a good idea is to have lots of ideas.—Linus Pauling
- Problems are only opportunities in work clothes.—Henry Kaiser
- Without creativity one creeps along far behind experience; with creativity one pushes ahead of it.—Edward deBuono
- Genius, in truth, means little more than the faculty of perceiving in an unhabitual way.—William James
- Originality is simply a fresh pair of eyes.—Woodrow Wilson
- Creative thought is adventure.—William Gordon
- Enthusiasm is the most important single factor toward making a person creative.—Robert Mueller

- The uncreative mind can spot wrong answers, but it takes a creative mind to spot wrong questions.—Anthony Jay
- The real mark of the creative person is that the unforseen problem is a joy and not a curse.—Norman Mackworth
- Creativity is not what is done but how one does it.—Barry Stevens
- It is axiomatic that to think intelligently is to think creatively.—Alex Osborn
- Almost every cloud does have a silver lining, but it takes imagination to find it.—Gordon MacLeod
- Daring ideas are like chessmen moved forward: they may be beaten, but they may start winning a game.—Goethe
- To raise new questions, new possibilities, to regard old problems from a new angle, requires creative imagination.—Albert Einstein
- Those who dare to fail miserably can achieve greatly.—Robert Kennedy
- The man with a new idea is a crank until the idea succeeds.—Mark Twain

Exercises

When searching for strategies (the "S" in DOC'S KEY), remember to focus on *quantity* of ideas, not on *quality* of ideas. There will be plenty of time, later on, to worry about quality. That's what the "K" in DOC'S KEY is all about.

1. Use the "brainwriting" technique (you'll need to team up with others) to generate at least twenty different strategies for how you might meet the following objective: "To increase membership in your favorite organization by 20 percent by the end of the year."

2. Using the "positive/negative forces analysis" technique, generate at least twenty strategies for how your organization might meet the following objective: "To reduce monthly energy consumption in your building by at least 8 percent by the end of the year."

3. Using "positive/negative forces analysis," generate at least twenty strategies for how a college's environmental curriculum could be improved.

4. You are a member of your community's planning commission (the group that oversees future growth of the town and reviews applications for

building permits). Using the "desired features" technique, generate at least twenty strategies for how your group might reduce the tension between developers, farmers, and land protection advocates.

5. Using "flowcharting" as an idea-generating technique, generate at least twenty strategies for how your household might reduce its waste stream by 10 percent by the end of the year.

6. Using "flowcharting" as an idea-generating technique, generate at least twenty strategies for how the following objective might be met: "Within the next three years, to slow by 10 percent the conversion of agricultural land to 'trophy house' development."

7. The popularity of "The Robert Frost Nature Trail" has grown by leaps and bounds, and some users are concerned about soil erosion and the spread of invasive, nonnative plant species. Using the "balloon diagram" as an idea-generating technique, generate at least twenty strategies for how these concerns might be alleviated.

8. Choose a volatile environmental issue (about which people feel strongly) and recruit six to eight people to role-play as different (but equally emotional) stakeholders. With you serving as facilitator, use the "nominal group technique" to generate at least twenty strategies for dealing with the problem.

9. Using "synonyms" as an idea-generating technique, generate at least twenty strategies for how urban sprawl might be abated.

10. Using "assumption reversal" as an idea-generating technique, generate at least twenty strategies for how the following objective might be met: "By the end of next year, to reduce annual roadside littering of beer bottles and cans by 30 percent."

11. Using "attribute restructuring" as an idea-generating technique, generate at least twenty strategies for how the following objective might be met: "By the end of the year, to increase weekly use of your community's recycling facility by 200 percent."

12. Using "exaggerated objectives" as an idea-generating technique, generate at least twenty strategies for how the following situation might be addressed: "NIMBY" (not in my backyard) does not seem to apply to Indian

reservations—their backyards are disproportionately targeted as toxic waste dumping sites.

13. Using any combination of idea-generating techniques, generate at least twenty strategies for how the following situation might be addressed: Attitudes about the ethical use of natural resources are shaped in large part by religious documents such as the Bible and Quran.

14. Revisit the important environmental problem that you first identified on page 1 of this book. Now turn to page 76 to revisit how you clarified your definition of this problem and associated goals, and turn to page 94 to revisit how you clarified your objectives and constraints. Transcribe these first three steps in DOC's KEY to the lines below so that you have them in one place and can inspect them to ensure that they're mutually compatible. Rerecording D, O and C will force you to refamiliarize yourself with each. As you get further into a problem, your take on the problem often shifts; when this occurs, adjust affected parts so that all elements in DOC's KEY are compatible. (So, if you find during transcription that one or more elements no longer quite fits, make adjustments so that what you record below represents your best effort at D, O, and C.

The Problem (as you have defined it most recently):

_____

_____

Your Goal:

_____

_____

Your S-M-A-R-T Objectives:

Objective I: _____

_____

Objective II: _____

_____

Objective III: _____

_____

Objective IV: _____

_____

Your Constraints:

_____

_____

_____

_____

_____

_____

_____

_____

_____

15. Revisit your "farmer's overview" from p. 94. Transcribe it (incorporating improvements that may come to mind) to the lines below.

_____

_____

_____

_____

_____

_____

_____

_____

_____

_____

_____

_____

_____

_____

_____

_____

_____

_____

_____

16. Revisit Objective I in 14 above. If it still seems appropriate, use techniques described in this chapter to generate at least twenty strategies for how this objective might be achieved. Record them on the lines below.

Twenty Possible Strategies for Achieving Objective I:

1. _____

2. _____

3. _____

4. _____

5. _____

6. _____

7. _____

8. _____

9. _____

10. _____

11. _____

12. _____

13. _____

14. _____

15. _____

16. _____

17. _____

18. _____

19. _____

20. _____

17. Revisit Objective II in 14 above. Using techniques described in this chapter, record below at least twenty strategies for how this objective might be achieved.

Twenty Possible Strategies for Achieving Objective II:

1. _____

2. _____

3. _____

4. _____

5. _____

6. _____

7. _____

8. _____

9. _____

10. _____

11. _____

12. _____

13. _____

14. _____

15. _____

16. _____

17. _____

18. _____

19. _____

20. _____

18. Revisit Objective III in 14 above. Using techniques described in this chapter, record below at least twenty strategies for how this objective might be achieved. (Experiment with techniques that you have not yet tried.)

Twenty Possible Strategies for Achieving Objective III:

1. _____
2. _____
3. _____
4. _____
5. _____
6. _____
7. _____
8. _____
9. _____
10. _____
11. _____
12. _____
13. _____
14. _____
15. _____
16. _____
17. _____
18. _____
19. _____
20. _____

19. Revisit Objective IV in 14 above. Using techniques described in this chapter, record below at least twenty strategies for how this objective might be achieved. (Experiment with techniques that you have not yet tried.)

Twenty Possible Strategies for Achieving Objective IV:

1. _____
2. _____
3. _____
4. _____

5. _____

6. _____

7. _____

8. _____

9. _____

10. _____

11. _____

12. _____

13. _____

14. _____

15. _____

16. _____

17. _____

18. _____

19. _____

20. _____

Chapter 8

How to Make Decisions That You Won't Regret Later: The "K" in DOC'S KEY

Hopefully, your idea-generation efforts produced promising strategies for each objective. However,

Before evaluating possible strategies for implementation, always add "do nothing" to your list of options.

In the final analysis, "doing nothing" may yield better consequences than the novel strategies that you are considering. Don't make things worse just for the sake of making them different!

With a range of possible strategies to choose from (including "do nothing") you're ready to make some decisions about which strategies to implement.

That is the focus of this chapter—evaluating and selecting the "keepers"—the "K" in DOC's KEY. If your group embraced the spirit of "anything goes", generating ideas and strategies was fun rather than stressful. This carefree mood may change abruptly for some team members, however, when the group confronts the challenge of selecting which ideas and strategies to implement. Other members of the team won't be the least bit fazed at committing to a decision.

How People Think and Make Decisions

People differ a great deal in how they think and make decisions. One of the differences revolves around the extent to which decisions are guided by "effortless" and "effortful" thought processes. Most people favor one more than the other.

Effortless decision making requires no output of mental energy—it's passive and automatic, and it's guided by habit or by things that you believe or think you know. Deciding to brush your teeth or deciding how to put on your socks are both examples of effortless thinking—without thought or consternation, you settle on particular courses of action and implement them without thinking twice about them.

Effortful decision making, in contrast, requires a conscious output of mental energy—it's active, measured, and it's guided by analysis or reason. You think "effortfully" any time you push yourself to analyze a situation from many angles before committing to a decision. Deciding how to manage your finances responsibly, for example, requires effortful thinking. Effortful decision making can be exhausting, so you probably avoid it whenever you can. Sometimes you probably avoid it when you shouldn't.

Effortless decision makers are often heralded as "decisive" because they make up their minds quickly, and they stick with their decisions. Dilly-dallying with details, or rethinking past decisions, is not their style. Effortless thinkers tend to favor simple solutions that have immediate payoffs; longer-term consequences can wait.

Effortful decision makers, on the other hand, actively engage in decision-making details that may seem unimportant to others. Effortful thinkers often appear indecisive because they are slow to make decisions—they seem to agonize over each new idea, detail, or longer-term implication. Resulting decisions tend to be complicated because all contingencies have been considered.

Environmental problem solvers want to be decisive, of course (unnecessary dithering annoys everyone), but not if the cost of decisiveness is effortless thinking.

When approaching environmental problems, decisions reached by effortless thinking lead to trouble—if not immediately, then later. Important decisions reached by effortful thinking take more time and energy, but they're more likely to be useful, fair, and enduring.

Effortless thinkers need to be directed into more effortful modes if they are to make meaningful contributions to a problem-solving effort. This can be promoted by phrasing problems, ideas, and possible strategies in unusual ways. Unfamiliar phrasing of problems breaks habit-bound, knee-jerk reactions and responses (i.e., effortless thinking) because responding by rote is no longer possible—there are no stock answers. Here is an example of how a challenge could be restated to encourage more effortful engagement: The challenge is, "How can we raise money for our organization?"

Trying to find funding for an organization is a very familiar problem, and effortless thinkers can easily and effortlessly respond by reaching into their bag of old, stale strategies—apply for grants, have a bake sale, etc. When a decision needs to be reached about which strategy to pursue, effortless thinkers always fall back on facile, familiar strategies.

The minds of effortless thinkers can be pushed into more effortful modes, however, by casting challenges and decision-making processes in less conventional ways. For example, "How can we raise money for our organization?" could be rephrased as, "How can we change our organization's financial situation?" Effortless thinkers now have no place to hide because the challenge is unfamiliar, with no stock answers. Everyone must think effortfully to contribute, so more thoughtful contributions result. When searching for the best strategies to an environmental problem, you need all the effortful thinking you can get.

Effortful thinking is needed to make good decisions about which strategies to pursue. That entails thinking effortfully about each of the two parts of decision making:
• evaluating the strengths and weaknesses of your possible strategies; and
• selecting which strategies you'll implement.
These need to be addressed one at a time.

The First Part of Decision Making: Evaluating Strengths and Weaknesses of Your Options

Anytime you evaluate possible options, you are looking, of course, for the "best" strategies to implement. But "best" is elusive, for any measure of goodness is based on your value system and the upbringing, culture, faith, economic status, and past experiences which have shaped you. "Best" is also dependent, of course, on what you judge to be the *real* problem.

Consider, for example, a wetland that borders the newly proposed elementary school playground. Some parents probably would consider the wetland a problem, for it produces mosquitoes that might bite their kids. To deal with the "mosquito problem," they would favor strategies that counteract the problem—draining the wetland, for example. It's very likely, however, that some town residents would view the wetland as an aesthetically pleasing buffer, or as a great place to go birdwatching. To these people, any proposed management strategy for the wetland would need to be evaluated in terms of how it will affect those values.

The outcome of decision making is predicated on which criteria you use to evaluate the strengths and weaknesses of your options.

So—your first challenge in evaluating possible strategies for implementation is deciding on the exact criteria by which each strategy will be evaluated. Given the wide range of values that interested parties hold, who decides which criteria to use as measuring sticks? It's tempting to stack the evaluative and decision-making decks with people who feel the same as you do. You then are pretty much assured of reaching the outcome that you personally favor. But deciding on a strategy is one thing, implementing it successfully is another:

If you want support for an environmental decision—which you almost always will if you want your decision to be implemented—you need to include stakeholders (interested parties) in all parts of the decision-making process, even if their views are very different from your own.

There is a cost to this: including diverse perspectives means that quick decisions will not be forthcoming.

It's rarely practical to include every single individual or group that's interested in an environmental problem, so you will need to make some tough decisions about inclusion and exclusion. This is likely to alienate those who are not included in the process because they will resent being left out. Expect these malcontents to dismiss whatever decision your group makes, no matter how good the decision might be. You can minimize this type of backlash, however, by inviting each relevant stakeholder group to select a representative to participate in the evaluation and selection process. The evaluation team is thus kept small in number but large in representation. This helps foster buy-in.

Once your group has been decided upon, your next challenge is to be upfront about what role participants will actually play in the evaluation and selection process. If the role of participants is only advisory (if you or someone else will make the final decision), be very clear about that before you begin. Letting participants harbor beliefs that they are equal players in the decision-making process (if they won't be) is a charade that will cost you dearly later on.

Evaluative Criteria for Possible Environmental Strategies

To decide which strategies to pursue, you first need to evaluate the strengths and weaknesses of each against agreed-upon evaluative criteria. Cost-benefit analysis (described below) is one evaluative approach that is commonly used to judge the relative strengths of different strategies. Few environmental problem solvers, however, are entirely comfortable with an approach that reduces all elements of an environmental problem to a single evaluative currency such as money. Other criteria probably also are needed.

A principal measure of the quality of any strategy should be, of course, the strategy's likelihood of meeting your objectives. Other measures of quality (evaluative criteria) also may be very important, however, and these mustn't be ignored or dismissed. Here are six evaluative criteria that are almost always relevant:

• How easily can the proposed strategy ("solution") be implemented?
• How quickly can the proposed strategy be implemented?
• How much will it cost to implement the proposed strategy?
• How politically problematic is implementation of the proposed strategy likely to be?
• How fail-safe is the proposed strategy likely to be?
• What environmental impacts might accompany the proposed strategy?

The last criterion, "environmental impact," has somewhat different meanings in different contexts. Within the National Environmental Policy Act

(NEPA) and Environmental Impact Assessment (EIA), its meaning is relatively specific. When associated with problems that don't involve the federal government in any way, its meaning is broader. Either way, however, environmental impact represents a condition against which possible strategies need to be evaluated.

There are three generally recognized types of "environmental impacts": direct, indirect, and cumulative. Direct environmental impacts are those that result directly and immediately from a project-related activity (e.g., a factory dumps chemicals into a river and fish die). Indirect environmental impacts are those that are offshoots of a direct impact (e.g., a factory dumps chemicals into a river, fish die, and tourism declines. In this example, decline in tourism is the indirect environmental impact). Cumulative environmental impacts are those that result from incremental but collective influences (e.g., a factory dumps chemicals into a river, the chemicals react with existing, benign pollutants in the river, and noxious fumes are created that peel paint off riverside houses). In this example, the peeling paint is the cumulative impact.[1]

All three types of environmental impacts (above) need to be considered when you are evaluating possible strategies for implementation. Do not discover the hard way that your "solution" (i.e., your implemented strategy) creates more problems than it solves.

Cost-Benefit Analysis

Cost-benefit analysis employs a single evaluative criterion. Strictly applied, cost-benefit analysis is used by economists to evaluate and compare different financial strategies, using money as the currency. The cost-benefit approach can also be used with other currencies such as time or energy consumption.

Cost-benefit analysis is an attractive evaluative technique because it reduces all considerations to a single, shared, comparable currency. Vastly different strategies can therefore be compared quantitatively. However, the devil is in the details. Getting people to agree on (or even to consider) the monetary worth of heartfelt considerations such as happiness, security, health, or extinction of a species is oftentimes impossible. The resulting

---

1. Peeling paint is considered a cumulative impact because, in isolation, the factory's chemicals would have no effect on riverside houses; only when combined with other, preexisting chemicals in the river does the impact (peeling paint) emerge.

stalemate can become enormously frustrating and divisive; something ultimately has to give. Proponents of cost-benefit analysis may wish to abandon others in the problem-solving group and go at it alone, but this would be a big mistake. In the end, there would be no buy-in from others and you'd be worse off than when you started.

If group members don't buy into cost-benefit analysis as the main evaluative criterion, don't push it. Treat cost-benefit analysis as one of several evaluative approaches.

Troubleshooting

Evaluating strengths and weaknesses of strategies vis-à-vis standard evaluative criteria is a good start, but almost certainly there are other factors that need to be taken into account. These additional factors need to be identified and added to the list of standard evaluative criteria. Troubleshooting is a way to do this.

Effective troubleshooting has three parts. The first entails drafting an expansive list of issues that might affect the problem-solving outcome. The second focuses on identifying which issues on your troubleshooting list are most likely to interfere with your problem-solving effort. The third centers on developing and implementing strategies to minimize their possible impacts. The second and third parts of troubleshooting depend, of course, on how well you ferret out issues that could disrupt your problem-solving effort. Two techniques, "talking out" (p. 63) and "brainwriting" (p. 98), are especially helpful at uncovering factors that should be considered.

Brainwriting

As discussed in an earlier chapter, brainwriting is a powerful tool used by think tanks to generate lots of different ideas in a hurry. Brainwriting works well in the first stage of troubleshooting (identifying issues that may affect how the problem is solved) because it invariably uncovers issues and evaluative criteria that have been overlooked.

To illustrate how brainwriting can be used to identify additional evaluative criteria, let's say that you are considering different strategies for dealing with the following problem: "The Franklin Forest is not being managed sustainably." After assembling a brainwriting group, reframe your problem as a troubleshooting challenge: "What things might we need to take into account to manage the Franklin Forest sustainably?" Now brainwrite for ten minutes and collect the brainwriting lists.

What to Do with the Collected Brainwriting Lists

The collected brainwriting lists hold many ideas, thoughts, issues, and possible evaluative criteria, but you can squeeze out a few more by "reverse troubleshooting." Begin by brainwriting: "What can or might go *well* if we implement some of these strategies?"

Your brainwriting lists now hold many important criteria, but they also hold many entries that are not useful. Now is the time to separate the wheat from the chaff. Do so by discarding senseless, irrelevant, duplicate, and off-the-wall entries. The resulting, streamlined brainwriting list represents criteria against which possible strategies might be evaluated. Combine this brainwriting list with the six evaluative criteria described earlier. The combined list of grouped thoughts, issues, and ideas becomes your "grand list" of evaluative criteria.

Distribute copies of the grand list to members of your problem-solving team so that each member can evaluate the importance of each item on the grand list. Ask team members to rate each item as a "3" (very important or likely to be very important), a "2" (fairly important or may be important), or a "1" (probably not very important). The rated grand lists are then collected, and scores are tallied for each item on the list.

Based on the tallied scores, possible evaluative criteria on the grand list can now be ordered according to their perceived importance. The ordering process usually reveals a few items that almost everyone thinks are very important, a sizable number of items that are judged to be fairly important, and a few items that almost everyone judges to be relatively unimportant.

Now solicit reactions to your ordering of grand list items by running your ordered grand list past people outside your problem-solving group. If meaningful feedback is slow in coming, prime the feedback pump by asking, "Do you agree with our assessment of which evaluative criteria are most important? Have we forgotten anything important?" Carefully consider the resulting feedback and incorporate it into a final, revised grand list of evaluative criteria.

Before proceeding further with the revised grand list, first look for discarded items that were identified by at least one evaluator as "very important." Often, one person sees something in an item that others do not. Seek clarification on why one person judged an item "very important" when others did not, then reevaluate the item's likely importance to determine if it should be added to the final revised grand list of evaluative criteria.

Your final revised grand list of evaluative criteria should now be extensive, but the relative importance of each criterion still needs some adjustment, for some criteria are certain to be much more important than others. These criteria need to be weighted to reflect their relative importance. Start the weighting process by identifying criteria that are so important that they *must* be satisfied for a strategy to be considered worthy of implementation.[2] Now assign an importance value (1–10: 1 = not very important, 10 = extremely important) to each of the remaining criteria. Use care when assigning an importance value to a criterion because the value you assign strongly affects the evaluation process.

Evaluating the Merit of Possible Strategies

Having made decisions about the relative importance of your evaluative criteria, it's now time to see how each strategy measures up. Begin by checking to see which of your possible strategies pass the first test of acceptability (i.e., that they *must* be satisfied for a strategy to be considered worthy of implementation). Strategies that fail this first critical standard should be removed from the pool of possible strategies.[3] Now subject the streamlined list of strategies to your other evaluative criteria (e.g., likelihood of succeeding, cost, staff needed to implement, time to implement) and numerically rate the extent to which each strategy meets each criterion. Multiply this rating by its assigned importance to yield a subtotal. Add subtotals for each possible strategy and compare the totals to see which strategies receive the highest scores (see table 8.1).

This weighting system provides a quantitative score for each strategy that you're considering and seems to remove the uncertainty of which strategies are actually best. But remember that quantitative scores are mean-

---

2. Strategies that fail this test of acceptability (e.g., the environmental impacts are unacceptable) are often discarded from further consideration. A wiser approach, however, is to search for ways to overcome their shortcomings. Versions that are more acceptable than the original strategy can then be considered for implementation.

3. Don't abandon these strategies completely, however. After completing an evaluation of the other possible strategies, return to your discarded strategies and brainstorm possible ways of overcoming their fatal flaw.

Table 8.1.

Evaluating Strategies Using Weighted Evaluation

Example of how different strategies can be evaluated and compared using weighted evaluation.

		Effectiveness of Three Strategies					
Evaluative Criteria	Importance	(1)		(2)		(3)	
• Likelihood of succeeding	9	×1	=**9**	×4	=**36**	×7	=**63**
• Inexpensive to implement	8	×9	=**72**	×1	=**8**	×9	=**72**
• Few staff needed to implement	4	×5	=**20**	×9	=**36**	×3	=**12**
• New members are activists	3	×1	=**3**	×8	=**24**	×7	=**21**
• New Members are well-healed	7	×2	=**14**	×6	=**42**	×1	=**7**
Totals			175		118		146

Notes:

Scale: 1 = low, 10 = high importance/effectiveness

Importance = how essential a criterion may be

Effectiveness = the likelihood that a strategy will satisfy a criterion

Numbers in bold = calculated subtotals (importance × effectiveness)

Strategies: (1) solicit by telephone; (2) purchase member lists from other NPOs; (3) target college outing clubs

ingful *only if* your assignments of importance have been accurate, and *if* you have included all relevant criteria by which possible strategies should be evaluated.

Best efforts notwithstanding, you oftentimes will have a gut feeling that the scores are not quite right, that the strategy with the highest score is not really the best. Do not dismiss your uneasiness prematurely; search for why you feel that way. It is likely that your instincts are right—that important evaluative criteria were forgotten or that your assigned values of importance to criteria were inaccurate. Make adjustments and recalculate totals for your possible strategies—not to give you the answer you want, but to reflect reality as you see it. Summary scores resulting from weighted evaluation are seductive because they appear to be fair, quantitative assessments of which strategies are best. Before swallowing summary scores as gospel, however, graphically profile each strategy that you are considering.

Graphical Profile

It's always wise to compare your best strategies side by side, criterion by criterion. The graphical profile does that. To create a graphical profile of the different strategies that you are considering, place the most important evaluative criterion at the top of the page, with the least important criterion placed at the bottom. For example, if speed of implementation is critical (e.g., a late solution is unacceptable), place "speed of implementation" first on your list. If speed of implementation is less important, but cost of implementation is absolutely crucial, place "cost of implementation" above the "speed" criterion. A profile of the merits of each alternative can then be graphed to illuminate differences among the possible strategies (fig. 8.1).

The graphical profile reveals strengths and weaknesses of each possible strategy, which makes comparison easier. The profile also identifies opportunities for improvement in your strategies. Might it be possible to create a new, composite strategy that steals the best features of other strategies? Oftentimes a new, composite strategy emerges as the best overall strategy.

The Second Part of Decision Making: Selecting Which Options to Implement

With the evaluative part of decision making behind you, it would seem that the best strategies for solving your environmental problem should now be quite obvious. Unfortunately, that's not the way it usually turns out. There are many variations on how to come to a decision, including some that are computer-assisted. In the end, however, you really only have three choices:
• make the decision yourself (executive decision),
• decide by vote, or
• reach consensus.
The strengths and weaknesses of these options are described below.

Executive Decision Making

Executive decision making is a top-down approach commonly used by CEOs and military leaders where the appointed leader single-handedly makes the decision and everyone else lives or dies by that decision. Executive decision making is simple, quick, transparent, unambiguous, and decisions are more likely to be daring or offbeat than those reached by vote or consensus.

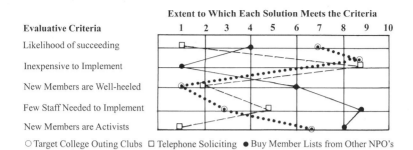

Fig. 8.1. Evaluating Strategies using a Graphical Profile
Example of how to graphically profile the extent to which various strategies are likely to meet important criteria (listed in order of importance).
Objective: Increase membership in your organization by 15 percent by the end of next year.

Most organizations recognize the need for some executive decision making (it's unproductive to hold a meeting every time you need to decide which brand of pencil to buy, for example), but:

Too much executive decision making is generally bad for group morale. When group members play no role in what is decided, they feel less invested in making the selected strategy succeed.

The exception to this is when a group is in crisis and doesn't know what to do. When this occurs, putting faith in a strong, respected leader to make the best decisions soothes frayed nerves. Be cautious about resorting to executive decisions, however, unless group members are strongly hinting that they want you to take charge. Usurping the decision-making process—when you shouldn't—will lead to trouble later on.

If you do resort to executive decision making, seek the wisdom of others to guide your decision. Tap the brains of friends, advisors, workers, representatives, confidantes, and those who hold views different from your own. Unofficial, anonymous, nonbinding votes (straw votes) can also be used to guide your decision making, Avoid publicizing the results of these votes, however. Group members will not take kindly to your overriding the group's will.

Be upfront with everyone, from the start, if you alone will be the one who makes the final decisions. When group members are unclear about their role in the decision-making process, many will wrongly assume that their opinion matters. When they discover it doesn't, group meltdown begins.

Deciding by Vote

Reaching decisions through voting is simple, straightforward, understandable to participants, and it spreads responsibility among participants.[4] A decision reached through voting engenders less buy-in than a decision that is reached by consensus, but usually more than a decision reached by executive decision.[5]

There are four common ways to reach a decision by vote:
• vote openly (by voice or show of hands),
• vote by secret ballot,
• vote by assigning points, or
• vote by decision matrix.

For routine matters where no one is likely to hold strong views, such as whether to accept the minutes from a prior meeting, an open vote works fine—it's quick, definitive, and voters are not likely to be influenced or threatened by status or personality differences within the group.

Open votes are ill-advised if you truly wish to gauge a group's preference on substantive matters, or if your group needs to make a quick decision on something of importance. Opt for secret ballot instead.[6] When people vote openly, they naturally shy away from positions that might be unpopular or might go against the leanings of a recognized leader or boss.

Voting by Assigning Points

This decision-making approach allows voters to express their level of enthusiasm for ideas and strategies. Give voters ten points each and instruct them to apportion their points as they see fit; whichever idea or strategy ultimately receives the highest total number of points from voters is the

---

4. No one single person therefore bears the entire burden if an implemented strategy does not work.

5. Note that mundane decisions are best made by a single person who has the group's best interests at heart. This also is the case for decisions made during times of crisis or extreme duress. Note also that "deciding by vote" is problematic when the interest of the individuals is different from what's best for the organization.

6. Voting by secret ballot may not result in a clear majority opinion. When this occurs, hold a second secret vote, but limit voting choices to options that were most popular in the first vote.

preferred option. For example, if one person thinks option C is by far the best option of those presented, that person might choose to allocate five or six of his/her points to option C. If option F seems a good idea also, but not as good as C, the person might give two or three points to option F. If option A has some merit, but less than option F, the person might choose to support it with a single point. By comparison, a different member of the group might allocate three points to option B, two points to option C, four points to option F, and one point to option G.[7]

Voting by Decision Matrix

This approach can be used for both individual and group decision making. Many problem solvers think that voting by "decision matrix" is the fairest and most thoughtful voting approach because the method forces pairwise comparisons of all options. When all pairwise comparisons of options have been voted on, the option receiving the most votes wins.

To illustrate, let's say that the State Wetlands Office wants help deciding how to manage Round Pond State Park. The office has identified ten wetland functions that are especially valuable and has sent the following questionnaire (a decision matrix) to a number of important stakeholders:

The following ten management options are being considered for Round Pond State Park:

____ a. Fish habitat

____ b. Surface and ground water protection

____ c. Water storage for flood water and storm runoff

____ d. (Non-fish) wildlife habitat

____ e. Biological diversity

____ f. Threatened/endangered species habitat

____ g. Education and research

____ h. Recreation

____ i. Open space/aesthetics

____ j. Water quality improvement

Please help us decide which options to emphasize by considering each pairwise comparison below, indicating which option (of each pair) you prefer.

7. As with other voting approaches, assign votes anonymously so that individuals are not swayed by those who vote before them.

Example:

ⓐ   b

a   ⓒ

Figure 8.2 shows one stakeholder's response.

When all comparisons have been made, tally the number of times each letter (i.e., management option) was circled and record the totals. The management option with the highest number of votes represents your first choice; the option with the second highest number of votes represents your second choice, etc. In the Round Pond example, recreation (option "h") was most favored by the stakeholder whose response is shown in fig. 8.2, with non-fish wildlife habitat (option "d") and open space/aesthetics (option "i") close behind. Following are this stakeholder's totals:

_3_ a. Fish habitat
_2_ b. Surface and ground water protection
_4_ c. Water storage for flood water and storm runoff
_7_ d. (Non-fish) wildlife habitat
_3_ e. Biological diversity
_2_ f. Threatened/endangered species habitat
_3_ g. Education and research
_9_ h. Recreation
_7_ i. Open space/aesthetics
_4_ j. Water quality improvement

The decision matrix is executed in the same way for groups as for individuals. Each voter privately completes the decision matrix. If a group decision is desired, individual tallies are combined to yield an overall score. The option with the most votes constitutes the group's decision.

Deciding by Consensus

Deciding by consensus means that everyone buys into the same decision, even if the decision is no one's first choice.

Environmental decisions reached by consensus tend to be more easily implemented, and more enduring, than those that are reached by other

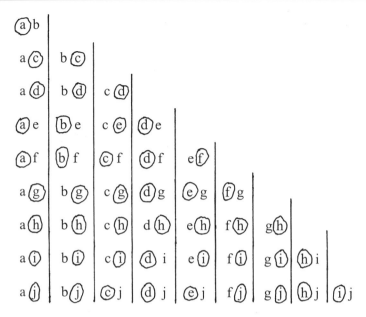

Fig. 8.2. Decision Matrix for Round Pond State Park: One Stakeholder's Response

means. Making decisions by reaching consensus can be agonizingly slow, however, and imaginative strategies are often whittled to a lowest common denominator. Some problem solvers complain that decisions reached by consensus lack backbone.

To speed the consensus building process (and to keep people from quitting out of frustration), put a timeline on the process. Do this by clearly and unambiguously informing participants that a decision *must* be reached by a specific, clearly designated time. Explicitly state what that time is. Let participants know that you are hoping to make the decision together as a group, but if consensus cannot be reached by the designated time, you will need to resort to voting or executive decision, two options which you find less desirable. Being upfront about time constraints will motivate participants to work together more cooperatively. It also will buffer you from snipings that you took things into your own hands.

To reach consensus, you need to work together cooperatively, even when your value systems differ. For example, if you're working with a diverse group to establish policy on how a forest might be managed, it's likely that some members of your group will feel that they *know* that clear-cutting forests is bad. It's also likely that some other members will feel that they *know* that clear-cutting is not bad. How is consensus possible in situations such as these where participants are certain that they are right and that others are wrong?

Getting to know one another on a personal level is a good start, for it softens black-and-white attitudes about right and wrong and good guys and bad guys. The words of Rudolf Flesch should be read often:

> We all pride ourselves on having an open mind. But what do we mean by that? More often than not, an open mind means that we stick to our opinions and let other people have theirs. This fills us with a pleasant sense of tolerance and lack of bias—but that isn't good enough. What we need is not so much an open mind—readiness to accept new ideas—but an attitude of distrust toward our own ideas.[8]

A shift in attitude in the direction of "open-mindedness" begins with acknowledging that the "truths" that you hold dear may not, in fact, be truths after all. Maybe, just maybe, some of your "truths" are instead viewpoints or attitudes that were shaped by your upbringing. Perhaps, if you had lived a different life in a different place or time, you would hold different, but equally strong views. Maybe those views would be as true as the ones you now hold.

Work to cultivate that thought process in yourself. (It's not easy!) When you and others in your group begin to shift your thinking in that direction, there is hope of reaching real consensus.

Finding Common Ground

Fostering a distrust of your own "truths" makes finding common ground easier.

"Common ground" is a perception of the problem that all of you can live with. It is the foundation of consensus building.

To move a group in the direction of finding common ground, begin by doing a quick round-robin, where each person articulates aloud (without interruption or challenge) what s/he perceives the real problem to be.[9] This gets everyone's ideas about the problem out on the table and exposes dif-

8. R.F. Flesch, *The Art of Clear Thinking*. New York: Harper and Bros., 1951: p. 176.

ferences in perspectives about the problem and the best strategy to solve it. This is a good, neutral starting place for trying to work together.[10]

Most problem-solving stalemates and ongoing battles result from opposing camps never getting beyond their deeply held views about strategies—the actions that should be taken to change the situation. When people are clinging to premade solutions and getting nowhere except increasingly frustrated with one another, request a timeout so that everyone can read through the first parts of chapters 4 and 5 (pp. 38–48 and 54–63) and get on the same, neutral page. Reconvene after participants have had time to cool down and digest what they've read.

After reconvening, help one another articulate clearly the exact situation that is perceived as unacceptable or undesirable—without passing judgment on whether or not you agree with one another's assessments. Use a second round-robin to do this. As participants have a heightened understanding of the *real* problem as others see it, the nature of the problem becomes less personal and less emotionally charged. People still may disagree strongly with one another, but the resulting discourse tends to focus more on the issue at hand, and less on showing that others are wrong. Doing a couple of round-robins, supported by the neutral guidelines (p. 9), is remarkably effective at helping people from opposing camps get beyond knee-jerk negativity to ideas that are not their own. Participants then can begin to hear what others are saying, and consensus becomes a possibility.

If people are committed to the effort, even warring factions can agree on some aspect of the problem and goal.[11] Once you all have reached that milestone, you can work on setting S-M-A-R-T objectives—ones that, if met, will constitute a solution to the part of the problem that you agreed upon.

Crafting objectives that are mutually agreeable is not easy, because opposing camps will naturally distrust one another's motives and agendas.

9. To protect against soliloquies and long-windedness, ask people to limit their comments to 45 seconds or less.

10. Round-robins often uncover strongly held beliefs that the presenter accepts as an inviolable "truth." Round-robins also uncover premade solutions that are posing as problems. Until the true nature of misidentified problems is understood by those who present them, consensus building will be next to impossible.

Self-awareness is the most powerful antidote to stilted thinking, so encourage fellow group members to read the first five chapters of this book. If that's asking too much, ask them to at least read the first few pages of chapters 4 and 5.

11. Remember: useful problem definitions are clear, impersonal descriptions of situations that you'd like to be different from what they are.

Stay focused on the agreed-upon problem and goal, be patient, and be on the lookout for mental relapses—yours as well as those of others. Eschew premade solutions that you have championed before; don't mistake them for S-M-A-R-T objectives.

Your group's first batch of objectives will run the gamut, so expect to be discouraged about the group ever agreeing on anything. Hang in there. Respectfully record every objective that is offered so that the group can review them, one by one, for clarification and tightening. Now search for common ground in this list of objectives—objectives that everyone can support.[12] If no such objectives are forthcoming, do another round-robin to see where things stand. Preface the round-robin with:

- "What can you live with?" and
- "What can you *not* live with?"

Knowing where everyone stands makes it easier to move forward because you'll know where the bottlenecks are. Since agreed-upon objectives are the easiest to work on, tackle them first. As your group makes headway on finding and committing to strategies for these least contentious objectives, group members will start trusting and respecting one another more. This will make it easier to tackle more challenging objectives later on.

Words from the Wise

- Be positive and upbeat about the consensus-building process. Consensus building is hard but it usually is worth the effort. Avoid negativity.
- Do not compete with others, and do not argue strongly for extreme positions. If you find this difficult, reread Flesch's quote (above).
- Evaluate possible strategies on paper, where the evaluative criteria and strengths and weaknesses of the strategies are easily seen.
- Implement quick, easy, affordable, and safe strategies first.
- Fondness for a certain strategy/solution does not mean that it is the best—or even a very good—strategy. Be careful of becoming too attached to an idea or strategy.
- Frequently do round-robins to see where people are in their thinking. Listen respectfully and without interruption to each person's input. If views are far apart, immediately follow the first round-robin with another. You will be amazed at how much ideas migrate toward one another with each succeeding round.

12. Modifying or combining objectives may be necessary to achieve this.

- Prolonged frustration and dead ends are common complaints when problem solvers seek consensus. When frustration sets in, get away from the problem for awhile. Spend time by yourself, attending to other, routine matters. Your subconscious will work on the problem for you (see also chapter 7).

- Remember to focus on meeting manageable objectives rather than on seeking one single, mega-solution to the problem. "Dividing and conquering" enables you to engage different specialists or sub-groups to work on different objectives. The end result is a collection of better, less painful, more manageable, more participatory, more effective strategies.

Exercises

1. The surging deer population in a nearby town is seen as a problem by many residents because (a) there has been a concomitant increase in the incidence of Lyme disease (a painful malady for humans, spread by deer ticks); and (b) plantings in suburban yards are being destroyed by deer browsing. You have been asked to assemble a team to solve this problem. Who (which stakeholder groups) would you want to include on your team?

2. Managers of nuclear power plants are looking for sites to dump their waste. You have been asked to evaluate the suitability of three sites: (a) Newark, New Jersey; (b) a small, poverty-stricken town in southern Mississippi; and (c) an Indian reservation in northern Montana. As problem-solving team leader, what perspectives/stakeholder groups would you want to include in the evaluation process?

3. What evaluative criteria should be used to compare the three potential dump sites?

4. How should the decision be made vis-à-vis dumping of nuclear waste? Who should make the final decision?

5. Using techniques described in this chapter, identify, rank and display the ten most important criteria that should be used to evaluate the "quality" of the dump sites, as judged by:
   a. a nuclear plant manager
   b. a community resident
   c. the community leader or mayor

6. To shed light on possible career paths, you've decided to take stock of which job qualities are most important to you. Using the technique "voting by assigning points," indicate the level of importance you attach to each of following job qualities. (You have 20 points to apportion).

a. job security
b. working in a community of interesting people
c. making lots of money
d. spending lots of time outside
e. doing something that makes the world a better place
f. being in charge (vs. being told what to do)
g. traveling lots
h. prestige
i. being able to live where you want to live
j. having a job that has no after-hours responsibility
k. being near family

7. Rank the eleven job qualities above, from most to least valued, using a decision matrix.

8. Describe a situation where you and others tried to reach a decision by consensus. Was the effort successful? Why or why not? What difficulties did you encounter? How did you (or could you have) overcome these difficulties? Explain.

9. Describe two cases of *effortful* thinking that you have witnessed recently. Be specific.

10. Describe two cases of *effortless* thinking that you have witnessed recently. Be specific.

11. Refamiliarize yourself with the environmental problem that you have been working on since page 1 of this book. As you recall, you generated strategies for each of your objectives in the last set of practice exercises. In this set of exercises, you'll make substantive headway on the problem-solving road map by deciding which strategies you should implement.

a. Since the decision-making process is based on how you assess the strengths and weaknesses of each strategy, you first must decide on a set of evaluative criteria against which each strategy can be evaluated. Using techniques described in this chapter, do that now: identify and rank the criteria by which each strategy should be judged.

b. Now revisit the twenty or so strategies that you generated to meet your first objective (p. 136). Using techniques described in this chapter, evaluate each strategy against the evaluative criteria you identified above (a).

c. Graphically display the strategies that seem most promising.

d. Using techniques described in this chapter, select the strategies that should be implemented.

12. Repeat (a), (b), (c), and (d) from 10 (above) for your second objective (p. 137).

13. Repeat (a), (b), (c), and (d) from 10 (above) for your third objective (p. 138).

Chapter 9

How to Make Good Strategies Better: The "E" and "Y" in DOC'S KEY

Deciding which strategies to implement (the "K" in DOC'S KEY and the focus of the last chapter) is a real milestone. Congratulations for hanging in there.

Reflecting back on the problem-solving road map we've followed (DOC'S KEY), Homer Simpson would point out that there are easier ways to go about trying to solve environmental problems. The easiest way, of course, would be to jump on a party-line bandwagon and push a predetermined "solution" until you or the other guy drops. That approach would be satisfying if you were the one still standing—until reality raised its ugly head and you saw that your hard fought battle didn't accomplish anything except to make future enemies.

But that's not your way of approaching environmental problems: disciplined thinking, not knee-jerk reaction, is how you get where you want to go. With the first five steps of DOC'S KEY behind you (the D, O, C, S, and K), only the "E" and "Y" remain. Those two last steps in the problem-solving road map—experiment and yes, implement!—are the focus of this chapter.

Experimentation (the "E" in DOC'S KEY)

The "E" in DOC'S KEY refers to "experiment with your strategies"—work out the bugs before unleashing them on the world. Two experimentation techniques ("troubleshooting" and "decision tree flowcharts") are especially effective at finding potential flaws in your chosen strategies.

Troubleshooting

Troubleshooting was introduced in the last chapter as a technique to identify evaluation criteria against which possible strategies could be judged. The technique also can be used to identify shortcomings in the individual strategies that you wish to implement.

To seek out issues that could adversely affect how well a strategy works at meeting its objective, brainwrite (p. 98) or brainstorm (p. 101) the following: "What things might we need to think or worry about when trying to implement this strategy? What could go wrong?"

As outlined in the last chapter, drafting an expansive list of issues that might affect how well a specific strategy might work is the first part of troubleshooting. The second part of troubleshooting focuses on identifying which issues on your troubleshooting list are most problematic or of greatest concern. The third part of troubleshooting centers on developing and implementing ways to minimize possible adverse impacts.

Knowing what could go wrong—before it does—enables you to short-circuit difficulties before they become problems in themselves.

Many environmental problem-solving nightmares could be averted with a little troubleshooting, so avoid inclinations to skip over troubleshooting when you are on a tight budget or schedule.

The tighter your budget or schedule, the more essential it is that you troubleshoot!

Decision Tree Flowcharts

Flowcharts work well at revealing potential difficulties that could arise when you try to implement a strategy. To use a flowchart to uncover compromising factors, sketch the logical progression of steps that the possible strategy is expected to entail. The resulting visualization tracks how the strategy will be implemented, and what may happen at each stage of implementation. A "decision tree" flowchart thus emerges that provides an "if this, then that" portrait of the range of consequences that can be expected if a particular strategy is implemented.

Creating multipathway flowcharts for each of your favored strategies allows you to modify or debug possible strategies before implementation so that the strategies are more effective. Projected outcomes (desirable or undesirable) can be backtracked on the flowchart to find which antecedent factor was responsible for creating the outcome or bottleneck. Factors that lead to undesirable outcomes can then be minimized, and factors that lead to desirable outcomes can be maximized. Critical Path Analysis (CPA) and the Program Evaluation Review Technique (PERT), discussed below, are two well-established flowchart procedures that work well at identifying key decision points and bottlenecks. CPA and PERT also are effective at assessing the accuracy and acceptability of timelines.

PERT and CPA

PERT (Program Evaluation Review Technique) and CPA (Critical Path Analysis, also called Critical Path Method or CPM) are flowchart approaches that are used to plan and evaluate complex, multidimensional undertakings. PERT and CPA are very important experimentation techniques because they identify—before implementation—parts of undertakings that are problematic or unacceptable. This advance warning allows environmental problem-solving teams to correct deficiencies and bypass unnecessary obstacles.

Like all flowchart techniques, PERT/CPA[1] provides a sequential picture ("network") of how key tasks ("activities") link together to produce an outcome. Expected times to complete each activity are included in the network so that the total time to complete an undertaking can be estimated and evaluated. If the projected time exceeds the desired time, critical bottlenecks in the network can be identified and adjusted to reduce the project completion time to a more acceptable timeline.

To illustrate how PERT/CPA is used to forecast the effectiveness of a problem-solving strategy, let's say that your objective is to determine the top five outdoor recreation preferences of five hundred adult residents in Richmond, Virginia, by April 15. After articulating constraints and then generating possible strategies, you decide to keep and implement the following strategy: *conduct a mail survey of Richmond residents.*

---

1. The differences between CPA and PERT are subtle, so the description herein does not attempt to distinguish between the two.

Before committing all your time and resources to conducting a mail survey, it's wise to "experiment" a bit to see if the proposed mail survey is indeed a good strategy. Do you have time to pull it off? How would you organize the survey? Who would do what? Are there obstacles that you should be aware of? The three-part PERT/CPA process—building the matrix, building the network, finding the critical path—can help you answer all of these questions.

Building the Matrix

Begin the PERT/CPA process by making a comprehensive list of all activities (tasks) that need to be completed in order for the strategy under consideration to be implemented. An effective way to do this would be to brainwrite: "What tasks and activities might need to happen to execute the Richmond mail survey?"[2]

After creating an expansive list, discard those activities that are not essential to carrying out your strategy of creating, delivering, and processing a mail survey. The activities that you deem essential are then listed in a matrix, under the header "Activity" (table 9.1). (The order of activities in the matrix is immaterial.) Each activity is then assigned a unique "ID Code" (e.g., a, b, c, . . . z, aa, ab, ac., etc.).

The next step in building the PERT/CPA matrix entails coming up with a reasonable time estimate for how long it will take to complete each individual activity. When making these time estimates, it's better to overestimate than underestimate the time required to complete each activity.

The last step in completing the PERT/CPA matrix focuses on antecedents—exactly which other activities *must* be completed before the activity in question can begin. Antecedents vary among activities, of course—you certainly couldn't mail questionnaires (d) before you print them (e), or decide what questions to ask (a), or decide who gets a questionnaire (m)—so all three of these activities (plus some others as well) would be antecedents for activity (d), mailing the questionnaires.

When assigning PERT/CPA antecedents for an activity, include *only* those activities that absolutely *must* be completed before that specific activity can begin.

2. Make the list as complete as possible.

Table 9.1.

Experimentation Using PERT/CPA

How to build a matrix to identify possible difficulties with implementing a proposed strategy (to conduct a mail survey of Richmond residents). The matrix represents the first stage of experimentation using PERT/CPA.

The Question to Ask: What tasks and activities might need to happen to execute the Richmond mail survey?

Activity	Time Estimate (days)	ID Code	Antecedents
Formulate questions	3	a	none
Test drive questions	5	b	a
Address envelopes	2	c	m,n
Mail questionnaires	1	d	a,b,c,e,f,g,h,l,m,n,o,p
Print questionnaires	1	e	a,b,f,n,o
Find $ for printing	12	f	none
Figure out incentives to get people to respond	3	g	none
Find $ for postage	10	h	none
Collect questionnaires	5	i	a,b,c,d,e,f,g,h,l,m,n,o,p
Analyze results	4	j	all except k
Report results	2	k	all
Get helpers to label & stuff envelopes	7	l	none
Decide who gets questionnaire	3	m	none
Decide # questionnaires to send	2	n	none
Edit questionnaire	2	o	a,b
Stuff envelopes	1	p	a,b,c,e,f,g,l,m,n,o

Building the Network

The PERT/CPA network is a flowchart of connected circles and lines (fig. 9.1). Each line represents a specific activity; each circle represents the beginning of one or more activities (and usually the completion of others). The PERT/CPA matrix (table 9.1) provides all the information you need to build a network.

If a project is very involved (e.g., putting an astronaut on the moon), the matrix becomes so enormous that building a network is best done by computer. For smaller matrices, building a network by hand is more desirable because the process of building the network offers insights that otherwise might be missed.

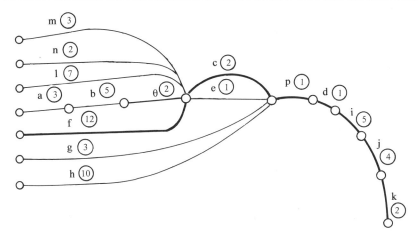

Fig. 9.1. PERT/CPA network for the Mail Survey Strategy
Each line in the network represents a specific activity, identified by a lowercase letter (see table 9.1). The circled number above each activity is the anticipated time needed to complete the activity (table 9.1). Each circle represents the start and/or end of one or more activities. There are a total of 12 different paths in the network, ranging in duration from 16 to 27 days. The bold line indicates the "critical path" (27 days)— the most time-consuming sequence of activities. *Note:* Length and shape of lines have no significance; some PERT/CPA formalities have been abandoned in this example.

Some people find it easiest to build a network by starting from the desired endpoint and working backward; others prefer to work from the beginning and move forward. Either way, the placement of each activity in the network is determined by its antecedents.

If you choose to work backward, begin by scanning your matrix for the activity or activities that cannot begin until all other activities have been completed.[3] If you choose instead to build your network in a forward direction, begin by scanning your matrix for the activity or activities that have no antecedents.[4]

Finding the Critical Path

"Critical path" refers to the sequence of activities that stands in the way of a strategy being implemented on a faster timetable. The critical path is enormously revealing because it pinpoints bottlenecks. Shortening the time it takes to implement a strategy can only be achieved by removing the

3. In the mail survey example, the final endpoint activity is "k" (report results) because every other activity is listed as an antecedent.
4. In the mail survey example, the starting activities are a, f, g, h, l, m, n because no other activities need to be completed before these activities can commence.

bottlenecks (i.e., by shortening the critical path). In the network shown in fig. 9.1 there are twelve distinct sequences (paths) of activities, with lengths of paths ranging from 16 to 27 days. Here are a few of them:

m→e→p→d→i→j→k (17 days)   m→c→p→d→i→j→k (18 days)

n→e→p→d→i→j→k (16 days)   a→b→o→c→p→d→i→j→k

(25 days)

The longest path (27 days) is the "critical path" because it is the sequence of activities that limits how quickly you can complete the mail survey. If 27 days is acceptable to you, then the mail survey strategy (as proposed) will meet your needs, and no modification is necessary. You can go ahead and implement the strategy as is. If the critical path is unacceptably long, however (i.e., you need to solve the survey problem in fewer than 27 days), you will need to find a way to speed up completion of one or more of the activities (f, c, p, d, i, j, k) that make up the critical path.[5]

Shortening the Critical Path

If the duration of the critical path exceeds your needs, you have a problem— the strategy under consideration will not give you the outcome you wish. Shortening the critical path then becomes a necessity. You can shorten the critical path in several ways:

• by omitting an activity in the critical path that is not absolutely essential;

• by replacing a lengthy activity with a less lengthy alternative;[6]

• by finding a way to hasten completion of one or more of the critical path activities.

Alternatives can be imagined through brainwriting or brainstorming. Simply ask, "In what ways might we shorten this critical path?"[7] If you cannot find a way to shorten the completion time enough to suit your needs (i.e., if the time to completion is still too long), you'll need to abandon the strategy and pursue others instead.

5. Expediting other activities (that are not part of the critical path) is wasted effort, for it will not shorten overall time to completion.

6. If delivery and return of the questionnaires by mail were activities in the critical path, for example, you could shorten the critical path by finding faster ways to deliver and collect the questionnaires. Possibilities might include delivering and collecting the questionnaires yourself, interviewing people in person, or conducting telephone surveys.

7. Remember to think wildly as well as sensibly—all ideas are good ideas!

Is PERT/CPA Worth the Effort?

The most challenging part of using PERT/CPA effectively is managing your attitude.

When a strategy that you're exploring is complicated, you're likely to find the process of building the PERT/CPA network frustrating and time-consuming. When a network doesn't come together quickly or smoothly, it's easy to start thinking that you're wasting time. Nothing could be further from the truth.

The process of building a network is as valuable as the network itself.

The PERT/CPA network represents a visualization of how you're thinking about solving the problem; the network is therefore a picture of your organizational plan for how your strategies will work. When parts of the network aren't fitting together or making sense, it's because parts of your strategy are faulty or incomplete. Effortful thinking and persistence in building the network will help you find and overcome these weaknesses.

Abandoning a network building effort when the going gets tough is tantamount to accepting a weakly developed strategy.

The Test Drive

Troubleshooting and PERT/CPA work well at strengthening your strategies and increasing the likelihood that the strategies will get you where you want to go. But some shortcomings in your strategies will probably still fall through the cracks.

Before committing to full-scale implementation, give your promising strategies a limited test run if at all possible. Often, the difference between "resounding success" and "bitter disappointment" is in the details, and some details won't be flushed into the open until your strategies are subjected to real-world forces. The test drive will reveal shortcomings in your strategies and give you a chance to make adjustments before launching them full throttle.

YES! Implement Your Strategies! (the "Y" in DOC'S KEY)

This is the moment you've been waiting for: YES! Implement your strategies! Meet your objectives! Solve the problem! If all goes as planned, your intimate relationship with this problem should now be over. Congratulations! Take a break!

But take a break with one eye open. After so much hard work, you don't want to let your guard down completely until you're entirely sure that your implemented strategies have gotten you where you want to go. If things start going slightly awry, deal with the situation sooner rather than later.

And don't forget the "undoers"; they're still out there. By dealing with everyone honestly, openly and respectfully, however, the undoers will be easier on you than they could be, and it will be harder for them to gain traction with others. Keep an eye on them nonetheless and respond to their scheming tricks directly and immediately, person-to-person. Don't let them drag you into a polarized "us versus them" dynamic. Abandoning the high ground will empower them.

Exercises

1. Let's say that a mandatory, citywide recycling program has been proposed to deal with the city's problem of an overflowing landfill. Choose a city with which you're familiar (that doesn't have a mandatory, citywide recycling program) and, using brainwriting as a troubleshooting technique, identify possible shortcomings, complications, and things that could go wrong with this strategy. (You'll need to team up with several others, of course, to do this).

2. Evaluate how much attention you should give to each of the brainwriting elements that your team generated for 1 (above). For each element, assign a score of "3" (a high chance of this element compromising your strategy's effectiveness); "2" (a moderate chance of this element compromising your strategy's effectiveness); or "1" (a low chance of this element compromising your strategy's effectiveness).

3. Create a PERT/CPA matrix for the recycling strategy (1 above).

4. Using the PERT/CPA matrix you just created for the recycling strategy (3 above), build a PERT/CPA network and find the critical path.

5. Revisit the environmental problem that's important to you (p. 1). Now revisit your selected strategies for meeting your problem objectives (p. 161). Using brainwriting as a troubleshooting technique, identify possible shortcomings, complications, and things that could go wrong with your *first* strategy. (You'll need to team up with several others, of course, to do this.)

6. Evaluate how much attention you should give to each of the brainwriting elements that your team generated for 5 (above). For each element, assign a score of "3" (a high chance of this element compromising your strategy's effectiveness); "2" (a moderate chance of this element compromising your strategy's effectiveness); or "1" (a low chance of this element compromising your strategy's effectiveness).

7. Using idea-generating techniques, strategize how you might circumvent or minimize the adverse effects of two of the most problematic elements. (Take the elements one at a time.)

8. Create a PERT/CPA matrix for your first strategy (5 above).

9. Using the PERT/CPA matrix you just created for your first strategy (8 above), build a PERT/CPA network and find the critical path.

10. Revisit your *second* strategy (p. 161) and use brainwriting as a troubleshooting technique to identify possible shortcomings, complications, and things that could go wrong with your *second* strategy.

11. Evaluate how much attention you should give to each of the brainwriting elements that your team generated for 10 (above). For each element, assign a score of "3" (a high chance of this element compromising your strategy's effectiveness); "2" (a moderate chance of this element compromising your strategy's effectiveness); or "1" (a low chance of this element compromising your strategy's effectiveness).

12. Using idea-generating techniques, strategize how you might circumvent or minimize the adverse effects of two of the most problematic elements. (Take the elements one at a time.)

13. Create a PERT/CPA matrix for your *second* strategy (10 above).

14. Using the PERT/CPA matrix you just created for your *second* strategy (13 above), build a PERT/CPA network and find the critical path.

15. After some fine-tuning prompted by troubleshooting and PERT/CPA, give your first and second strategies a test drive to see how they work. Identify weaknesses or flaws that would compromise the effectiveness of these strategies if they were implemented fully. How might you neutralize those weaknesses or flaws?

Chapter 10

Another Look at Problem-Solving Success and Failure

Earlier chapters presented environmental problem solving in terms of people factors and problem-solving steps, held together by the structure of "DOC's KEY." In this chapter, we'll look at five environmental problem-solving efforts to see where things went right and wrong.[1]

As stressed repeatedly throughout this book, environmental problems are rarely solved when an individual works in isolation, and they're rarely solved when the "people factor" is not addressed appropriately. It's not possible to describe on paper the full range of human interactions that are part of an environmental problem-solving effort, however. For that reason, the vignettes below instead call attention to particularly important successes or failings; they do not attempt to explore all aspects of "the people factor."

Five Environmental Problems and Their "Solutions"

"Finding Common Ground": An Environmental Management Strategy

The Situation: The rangeland conflict of the arid West was described in the first chapter, as were three strategies for solving it.

Analysis: Two of the strategies proved to be preordained solutions posing as "problems"; both were polarizing and ineffective. Neither was attentive to "the people factor," neither was the result of disciplined thinking, and

---

1. See bibliography for additional, more detailed case studies.

neither strategy was arrived at through an understandable approach such as DOC'S KEY.

The "common ground" strategy, in contrast, emphasized working *with* people rather than against them. The common ground strategy was a reasoned response to a well-defined problem, goal, and objective. It was not a preordained "solution" to an ill-defined problem. Last, the common ground strategy was spawned by an orderly, apolitical, and transparent sequence of problem-solving steps that was understandable to all parties.

Collecting Baseline Data: An Environmental Research Strategy

The Situation: "Baseline data" exist in many different forms, from observations recorded in a field journal (e.g., "two bears spotted along Sandy Creek") to quantitative measurements recorded on spreadsheets. The need for "baseline data" is a recurring theme in conservation organizations, and many organizations expend lots of money and effort trying to collect it.

Analysis: It would seem that you can't go too far wrong in collecting baseline data because *any* baseline data is better than *no* baseline data. But hold your horses—such effortless thinking is a thing of your past. You now think effortfully and, at minimum, ask:

• What problem is the "collect baseline data" strategy trying to solve? What is the goal? (the "D" in DOC'S KEY)
• What are the objectives? (the "O" in DOC'S KEY)
• What are the constraints? (the "C" in DOC'S KEY)

Environmental practitioners tend to bypass the "D" in DOC'S KEY when the solution to a problem is obvious, as in the case of collecting "baseline data." But the predetermined "collect baseline data strategy" is far from being an obvious solution because there's an infinite variety of baseline data out there, some of which is useful, some of which is not. Consider: Would you advocate collecting baseline data on the number of grains of sand on a wildlife refuge, for example? Or collecting baseline data on how many dandelions are flowering on the Commissioner's lawn?

Collecting baseline data is a predetermined solution looking for a problem. The presumed value of this strategy is based on assumptions—all of which usually prove to be faulty. For example:

- that people who come after you will understand and appreciate the value of your baseline data (Don't count on it!);
- that carefully collected baseline data will (eventually) reveal meaningful trends (Don't count on it! How, when, and where the baseline data are collected determines their usefulness later on.);
- that following "accepted procedures" will ensure that your baseline data have lasting value (Don't count on it! "Accepted procedures" for one problem won't necessarily be appropriate for a slightly different problem.);
- that the collected baseline data will actually be used (Don't count on it! Collecting data is easy; figuring out what to do with it after the fact is not. There's a reason why most "baseline data" eventually wind up in a dumpster!)

A better approach is to figure out what you're trying to figure out—*before* you go on a data collecting rampage! Be clear: what exactly is the problem? What exactly is your goal?

You might think that you can't really answer those questions. In fact, you might try arguing that the very reason for "collecting baseline data" is that "almost nothing is known about the place" (or) "there's no telling what might be important." Don't give in to such mental laziness: The truth is: (1) much more is known about the place than you're recognizing (you just haven't thought it through); and (2) you've already narrowed down the range of potentially important things by discarding those that you're sure are *not* important.

Consider the following:
- Let's get baseline data on how many people walk through the property each year in red sneakers.
- Let's get baseline data on what species live on the property.

The first one is ridiculous—who possibly could care? Who would be so dumb as to waste time and money on such useless data? The second one is more like it! Who could argue that getting this baseline information would be a waste of time or money?

Well . . . imagine that you're creating a management plan for the property and you have the "baseline data on what species live on the property" before you. Will having this species list help you make good decisions about what to do with different parts of the property? Mightn't you need to know other things, such as:
- Where on the property are they living?

• When do they live there?
• What are they doing there?
• How many of them are there?

To make management decisions for a property, questions such as these may be critical. Don't rush off to collect "baseline data" until you have asked and answered such questions.

Remember that:

"Baseline data" are useful *if and only if* they fit the problem you're trying to solve. (And that means you first have to figure out what the problem really is!)

and:

"Collecting baseline data" is a strategy—a means to an end—it's not an end in itself. Your research problem and desired outcome should always drive your strategy, not the other way around.

To state it yet another way, take care of the "D," "O," and "C" in DOC's KEY before tackling the "S."

Place-Based Landscape Assessment and Community Education:
An Environmental Activism Strategy

The Situation: You're troubled by a series of events in your town: prime agricultural land is being swallowed up by strip malls, wildlife habitat is being sacrificed for housing developments, and entire hillside forests are being clear-cut and auctioned off to out-of-state bidders. It doesn't seem that anyone cares.

Analysis: One strategy to slow down these land-use changes is to take legal action—tie up developers in court. Is this a winning strategy, one you should keep and implement?

The legal action strategy is often embraced as a "keeper" (the "K" in DOC's KEY), but is it a good choice? Were appropriate evaluative criteria used?

To answer that question you must look at the objective you're trying to meet. For example, if today's date is June 15 and your objective is "to halt all development of agricultural land by July 15," then the legal confrontation strategy might be your best or only strategy. If your timeline is less pressing, however, then it would be prudent to consider the consequences of ignoring "the people factor." Taking people to court is a surefire way to antagonize them and create an "us vs. them" dynamic. Once that line is crossed, you can forget about other, more cooperative strategies.

But let's say that your objective is somewhat different: "by December 1, to have at least 10 percent of your town's large landowners (those having more than 20 acres) participate in mapping out the town's natural resources." This objective (with associated constraints) would suggest a different set of possible strategies and a different set of evaluative criteria.

One "keeper" of a strategy for this objective is the PLACE program (Place-Based Landscape Assessment and Community Education).[2] This strategy emphasizes community participation in learning about the town's resources and is based on a tested premise that shared understanding and appreciation of the community's resources leads to thoughtful land-use policy and planning. PLACE brings people together, litigation pushes them apart. Think about it.

The National Environmental Policy Act: An Environmental Policy/Legislation Strategy

The Situation: As concern over the health of our environment grew in the 1960s, the federal government decided that legislation was needed to protect the environment from unacceptable impacts. The resulting federal legislation, the National Environmental Policy Act (NEPA), mandated that any proposed land-use having links to the federal government be assessed for adverse environmental impacts. The protocol for assessing the possible impacts (spelled out in NEPA) includes provisions for public review and comment.

Analysis: NEPA is a strategy designed to address the problem that "environmental impacts of proposed land-uses are not being assessed adequately." The protocol for assessment (legislated by NEPA) is built around a well-defined sequence of problem-solving steps that parallels DOC's KEY.

Solving the problem that environmental impacts of proposed land-uses

---

2. To learn more about this strategy and how it's implemented, check the PLACE website: http://www.uvm.edu/place/.

are not being adequately assessed is a complex undertaking, and the NEPA strategy is an impressive achievement. That's the favorable news. The unfavorable news is that *predicting* adverse environmental impacts is not equivalent to *stopping* them. To meet the latter goal, a different set of objectives and strategies is needed, subject to a different set of constraints.

Legislators recognized that "assessing the environmental acceptability of a land-use action" is very different from "using assessment results as the basis for deciding whether or not a land-use action is permitted." To address the decision-making aspect of the situation, they developed a separate strategy, in the form of a companion bill to NEPA. This companion bill never was voted on. And so:

• An organization (with links to the federal government) wishing to take an action that might have an adverse environmental impact is required by law to follow a specific protocol for how the potential environmental impacts are assessed.

• An organization is not legally required to comply with assessment findings. It's perfectly legal to ignore NEPA findings altogether, even if the action was determined to be environmentally disastrous.

Clearly, the goal—to stop adverse environmental impacts—was not met. The problem-solving effort lost traction because legislators were not sufficiently attentive to constraints—the "C" in DOC's KEY. They did a good job of defining the problem and goal, but their objective lacked an appropriate timeline (it was "S-M-A-R" but not "S-M-A-R-T"). The legislative session ended before legislators were able to reach a decision (pass the bill), and when the legislature reconvened, the political landscape (the people factor) had changed and the opportunity to pass the companion piece of legislation was lost.

The Field Naturalist Program: A Conservation Leadership Strategy

The Situation: By the late 1970s, some environmental leaders were becoming nervous about the near total emphasis on scientific specialization. The concern was that environmental problems span a number of interacting disciplines and that a collection of specialists from different fields is a weak substitute for integrative, big picture thinking. A related concern was the increasing disconnect between scientific understanding and environmental decision making: scientific understanding was not being made accessible to environmental planners and policy makers, and environmental decision making was suffering as a result. As a strategy to address these concerns, the A. W. and R. K. Mellon Foundations provided funding to several universities for the purpose of creating new graduate training programs. The University of Vermont, which proposed a "Field Naturalist Graduate Program," was one of those universities.

Analysis: The Field Naturalist Program has flourished where the other Mellon-funded programs have not because the founders of the program took the "D," "O," or "C" of DOC's KEY very seriously. For example, they worked through some very tough "D" (problem definition) questions:

- What is the *real* purpose of this program? What exactly is missing in current environmental training programs?
- When a graduated field naturalist walks out the door, what precise skills, talents and abilities do we want that person to have?
- What exact outcomes would need to be manifest for us to believe that the Field Naturalist Program is achieving what we'd like it to achieve?

Only after ascertaining exactly what they wanted to achieve did the founders move on to looking for ways to get there. Promising strategies (the "S" in DOC's KEY) were evaluated and decided upon (the "K" in DOC's KEY), experimented with and modified ("E"), and then implemented ("Y!").

But designing the Field Naturalist Program did not end there. Times have changed since the program was established in the early 1980s, as have problems and associated goals and objectives. Today's Field Naturalist Program continues to emphasize "hard, interdisciplinary science in the field," "oral and written communication," "integrated landscape analysis," and "conservation leadership," as it always has. But the elements of each of those four foci have changed over the years—sometimes subtly, sometimes dramatically—and objectives and strategies to meet those objectives have changed accordingly.[3]

Summary

The five environmental problems and strategies just described highlight common tipping points. You can tip the balance in your favor by breaking

3. More information about the Field Naturalist Program can be found on the program's website: http://www.uvm.edu/~fntrlst.

mind-numbing environmental problems into smaller, more tangible sub-problems, and crafting a goal (desired outcome) for each. As you know, you'd then tackle one subproblem at a time, focusing first on the *real* issue behind each subproblem (the "D" in DOC's KEY) and its subgoal. The next step would be to break each subgoal into specific, measurable, attainable, reasonable objectives (the "O" in DOC's KEY), identify constraints (the "C" in DOC's KEY), and so on until the subproblem has been solved.

You then could direct your attention to solving another subproblem, and then another, and then another until all subproblems have been solved to your satisfaction. Through this process of "dividing and conquering" you incrementally solve environmental problems that appear to be unsolvable.

Chapter 11

Go Forth and Move the World! Part I

This book (your problem-solving coach) has guided you through a training regimen of disciplined thinking and "people factor" awareness. Working with the coach, you've made considerable headway on the problem you first identified on page 1. The coach says you're now ready to get out there and do more. Show us what you can do!

But what *should you do?* With so many environmental problems to choose from, how do you know where to put your energy? And how do you fit "environmental problem solving" into the rest of your life? Is there a way to move the world and still put food on the table?

Putting together life's pieces in a way that works for you bleeds into an even scarier, more existential question: "What do you want to do when you grow up?" These questions are every bit as relevant and scary for an eighty-year-old as they are for an eighteen-year-old.

How To Solve This Problem

"What to do with your life" is a tough, unstructured problem where there's no premade formula for finding the right answer. That's the bad news. The good news is that, because the answer to this problem is far more important to you than to anyone else, you can use any problem-solving approach that you believe in—intuition, avoidance, emotion, faith, or disciplined, rational analysis. The opinions and reactions of family, teachers, and friends may matter to you a great deal, of course, but remember: *You* are the one who ultimately has to live with what you decide. Making a decision to please others might get people off your back in the short term, but "solutions" based on pleasing someone else quickly turn into cauldrons of regret and resentment. Stick up for yourself and what *you* want to do!

But that's just the problem: *you don't know what you want to do!* If you'd like to change that situation, might we suggest you give DOC'S KEY a try?

Step 1: Find and Define the Real Problem (the "D" in DOC'S KEY)

At first glance, the problem seems obvious: you don't know what to do about your life. Your goal also seems straightforward: to know what to do with your life. Past problem-solving experiences have convinced you to be leery of problem definitions that come too easily or too quickly, however. And so, even though you wonder if it's worth the time and effort, you dutifully scrutinize the problem using several problem-solving techniques—talking it out, "repeat why," reality checking, walking in other people's shoes, and freewriting.

As always, the search for the *real* problem proves to be well worth the effort, for it helps you better understand what you're up against. As it turns out, the situation that's bothering you may be two different subproblems:
• you don't know what you want to do with your life (and that's partly because)
• you don't know what your options are.
The corresponding subgoals would be (1) to know what you want to do with your life and, (2) to know what your options are. Solving the first (main) problem will be easier if you solve the second problem first, so we'll work on the latter one first.

Step 2: Set Objectives (the "O" in DOC'S KEY)

The two subgoals above codify and provide benchmarks for what needs to occur for you to feel that your problem has been solved. They don't tell you how to get there, however. That's the province of "strategies," the "S" in DOC'S KEY.

But not so fast. As anxious as you are to find an answer to your problem, you recognize that you first must break each subgoal into objectives. The resulting objectives will then provide you with specific targets to shoot for when you're developing and evaluating strategies later on in the problem-solving process. Setting targets greatly increases the likelihood that you'll get the results you ultimately want.

As you reshape your subgoals into more discrete objectives, you focus on making each objective "S-M-A-R-T": Specific, Measurable, Attainable, Reasonable, and with a designated Timeline. You also check to be sure that each stated objective has an action word (a verb).

Here are the S-M-A-R-T objectives that you might have come up with for the second subgoal (to know what your options are):

I. To identify, by March 1 of this year, at least ten different places, people, or resources from which you could get useful advice about how to look for or find promising opportunities.

II. To identify, by April 15, at least twenty different types of opportunities that might interest you.

Here are the S-M-A-R-T objectives that you might have come up with for the first subgoal (to know what you want to do with your life):

III. To find and read at least one "how-to" book on careers and life pursuits by May 1 of this year.

IV. By June 10, to be able to articulate (convincingly and specifically) at least one life pursuit that you'd jump on if you could.

V. To craft a detailed plan, by June 30, of what you'd need to do to have a realistic chance of making your desired life pursuit (IV above) come true.

Step 3: Ferret Out Constraints, Limitations, and Hidden Assumptions (the "C" in DOC'S KEY)

With the "D" and "O" from DOC's KEY in place for each subproblem, you're making great headway. Finding constraints, limitations, and hidden assumptions (the "C" in DOC's KEY) underlying each subproblem and subgoal is the next step in your guided problem-solving quest. Taking one subproblem/subgoal at a time, you begin by creating a skeleton profile of its boundaries and hidden assumptions through your responses to the following questions:

• Who is part of the problem?
• Who is not part of the problem?
• Who needs to participate in the decision-making process?
• Who should not participate in the decision-making process?
• When must a solution be available for implementation?
• For what time period must the solution be effective?
• What financial constraints are relevant?

- How much time do you really have to devote to solving the problem?
- What are the physical boundaries of the problem (where does the problem begin and end)?
- What legal considerations limit which problem resolutions are possible?
- What political factors limit which problem resolutions are possible?
- What environmental factors limit which problem resolutions are possible?
- What unacceptable impacts limit which problem resolutions might be viable?
- What other limitations, boundaries, and constraints define the scope of the problem?
- What assumptions are you making?

After taking this first cut at your problem's boundaries, limitations, constraints, and assumptions, you further explore real and imagined boundaries with the "5 W's and H" technique. This uncovers a large number of assumptions, many of which are shaky. Here are a few assumptions that have little or no grounding in reality:

- that you need to identify and settle into your life career right away;
- that you need to earn a certain income to be happy;
- that if you don't go to graduate school now, you probably never will; and
- That you can't travel and be a bum for a while.

Faulty assumptions (i.e., presumed limitations), such as those above, severely curtail your problem-solving options: Who says you *must* identify and settle into a career right away? Who says you *must* have a certain income to be happy? Who says you *can't* go to graduate school another time if you don't now? Who says you *can't* take some time off now to travel and be a bum?

It's easy to buy into these "musts" and "cant's" because you think you *need* to make money, and you think you *have to* start paying off loans, and you think you *can't* let your family down, and you think you *mustn't* burn bridges. But many of these are assumptions, not absolute truths. For example, who says you *need* to make money? Is it *money* you need, or is it the services and lifestyle and prestige that money provides? If you were presented with an opportunity where you made little money but had everything you wanted—car, toys, food, travel, a great community and place to live, health insurance, a chance to do something exciting and meaningful for the environment—would you dismiss the opportunity as unacceptable?

Flushing into the open real and imagined boundaries and assumptions is liberating because it separates reality from assumption. As you launch into the next step (looking for strategies), you're less likely to limit your options unnecessarily, and you're more likely to generate creative strategies that get you the outcome you most desire. You're also less likely to waste time trying to solve problems that are imagined rather than real.

Step 4: Craft Lots of Strategies for Each Objective (the "S" in DOC'S KEY)

Looking for ways to meet your objectives is the fun part because you can (and should!) let your imagination run wild. Good, bad, crazy, sensible, offbeat, and conventional strategies are all fair game—anything and everything goes. Knowing that the best way to come up with great ideas is to come up with lots of ideas, you take active measures to maximize your creative thinking. This begins with placing a moratorium on judgment and evaluation of your ideas.

You use several techniques (e.g., brainwriting, positive/negative forces analysis, flowcharting, discussing your problem with people outside your circle) to generate lots of ideas. Here's a sampling of some strategies you generated to meet your first objective ("to identify, by March 1 of this year, at least ten different places, people, or resources from which you could get useful advice about how to look for or find promising opportunities"):

- Ask anyone and everyone for advice on where to look.
- Search the web for promising leads.
- Identify people who seem genuinely happy—how did they figure out how to run their lives?
- Seek advice from a clergy member.
- Seek advice from those who know you best.
- Pray.
- Go to prison so you have time to think.
- Ask a reference librarian for help (they always seem to know where to look).
- Check in with career counseling.
- Check the yellow pages for ideas.
- Look for opportunities in as many different types of publications as you can.
- Find people who are successful at what you like or want to do—ask them for guidance.

- Check employment pages of magazines and newspapers.
- Study bulletin boards that are in places where you like to hang out.
- Schedule meetings with your favorite professors and ask them for help.
- Check in with alums who graduated with your major—seek their advice.
- Decide where you want to live and focus your search there.
- Find out which organizations and businesses do the sorts of things that interest you—learn more about those places on the web or elsewhere.
- Once you've identified an organization or business that seems to match your own interests, seek out someone from that organization or business to offer you insights and advice.
- Don't worry about it (do nothing). Let things happen as they may.

With a broad range of strategies to choose from, you now decide which ones to keep and implement. That's the next step in DOC'S KEY.

Step 5: Select the "Keepers"—Strategies to Implement
(the "K" in DOC'S KEY)

To make good decisions about which strategies to implement or discard, you first need to assess their relative strengths. You begin this assessment process by measuring each idea against six standard evaluative criteria:

- How easily can the proposed strategy (idea) be implemented?
- How quickly can the proposed strategy be implemented?
- How much will it cost to implement the proposed strategy?
- How politically problematic is implementation of the proposed strategy likely to be?
- How fail-safe is the proposed strategy likely to be?
- What environmental impacts might accompany the proposed strategy?

For this particular objective ("To identify, by March 1 of this year, at least ten different places, people, or resources from which you could get useful advice about how to look for or find promising opportunities."), some of the standard evaluative criteria are more relevant than others.[1] As it turns out, none of your strategies appears to be costly, politically problematic, or environmentally questionable. Ease and speed of implementation and likelihood of success surface as the most important criteria.

A number of your strategies appear promising and, when trouble-shooting fails to reveal any hidden obstacles or concerns, you decide to im-

---

1. To keep things simple, let's say that your search for other evaluative criteria that may be important didn't turn up anything new.

plement several of them. Time management then becomes the main consideration, so you piece together a timeline or flowchart to guide how and when you implement the chosen ideas. The resulting timeline (which functions as a reality check) confirms that it's very possible to pursue several strategies at once and still meet your objectives by the stated timeline. You therefore decide to implement all of them as planned. Bingo, you've met your first objective.

Successfully meeting your first objective now positions you to meet your second objective ("To identify, by April 15, at least twenty different types of opportunities that might interest you."). To meet this objective, you only need review the possibilities that you're now aware of (that were revealed when you met your first objective) and then single out those that interest you.

Solving the Other Subproblem: "What Do You Do with Your Life?"

"Not knowing what to do with your life" is often portrayed as life's biggest challenge because so much is on the line—the decisions you make now set the course for how your life will go. Or so you think. But is it necessarily so? Are you making assumptions that aren't necessarily true? For example, *Who says* you must set in stone your life course? *What* makes you think that you should know—today—what you'll want to be doing twenty years from now? *Why* can't you, instead, think of your life as a collection of chapters, with each chapter being an adventure unto itself? *Why* do you think you need to figure out your last life chapter before you figure out the chapters that precede it?

It's always wise to revisit the accuracy of a problem definition (the "D" in DOC's KEY) before moving on to subsequent steps. That's what we did in the preceding paragraph. By employing the "Who? What? Where? When? Why? How?" (5Ws and H) technique, we reevaluated the accuracy of the previously stated problem (that you don't know what you want to do with your life). It may be that your efforts to define the problem anew don't change a thing about your perception of the problem (i.e., you're still troubled by not knowing what to do with your life). It may also be, however, that the reexamination of your problem leads you to different understandings of the "problem." For example, the redefinition process may cause you to realize that there's no problem to solve because, "Hey, you *like* not knowing what's ahead of you! Life's more of a day-to-day adventure that way!" It also could be that it's a family member, not you, who is troubled by uncertainty in your life. If that's the case, then it's *that person's* problem, not yours.[2]

To keep things simple, however, let's say that the original problem definition (you don't know what you want to do with your life) continues to describe a situation that you'd like to be different. Here's how you'd solve that problem using DOC's KEY.

You've already completed the first two steps ("D" and "O") in DOC's KEY and you've made considerable headway on the third step ("C") as well. There probably are additional constraints that impinge on meeting one or more of your objectives, however, and these need to be ferreted out.

Meeting your third objective ("To find and read at least two "how-to" books on careers and life pursuits by May 1 of this year.") is a straightforward strategy, so we'll move on to your more challenging fourth objective: "By June 10, to be able to articulate (convincingly and specifically) at least one life pursuit that you'd jump on if you could."

To meet this objective, you need to know the elements that collectively make a life pursuit seem great to you. But what are those elements? You surely know, or think you know, some of the elements that you most value, but do you know them all? Apparently not, or you wouldn't be trying to meet this objective!

To meet this objective, you need to meet another couple of objectives first:
• To identify, by June 1, the ten elements that are most important to you in defining "a great pursuit."
• By June 3, to rank those ten elements in terms of their importance to you in a life pursuit.

This turn of events (adding another layer of objectives to what seemed a pretty straightforward problem-solving effort), seems to make solving the problem harder, not easier. But you've been through this before and trust in the value of "dividing and conquering." You know that DOC's KEY will keep you on track so that you don't lose your way.

With the focus now on the two new objectives above, you move on to the next problem-solving step: finding constraints, assumptions, limita-

---

2. Unless the family member's anxieties are unacceptable to you. In that case, your problem and goal would revolve around the family member no longer having those anxieties.

tions, and boundaries for each objective (the "C" in DOC's KEY). Because each objective centers on finding your own personal preferences and values, however, it turns out there are no constraints—anything goes if it suits your fancy.

"Finding Strategies" (the "S" in DOC's KEY) is your next order of business and you come up with many different approaches. On evaluating their relative strengths (the "K" in DOC's KEY), two strategies seem especially promising, and those are the strategies that you decide to keep and implement:

1st strategy: Find and take a "quiz" that reveals personal interests and values.

2nd strategy: Find real-world opportunities that excite you, and identify the elements that make them exciting.

With respect to the first strategy, you probably won't find many "environmental interests" quizzes out there, so the following quiz has been prepared for you. Take the quiz now to reveal your interests and preferences.

The "What do you value?" quiz

Consider each of the following and evaluate its importance to you. Circle each element that is very important to you; place a "$\sqrt{}$" beside each element that is somewhat important to you; ignore elements that are not important to you.

a. ___ to spend lots of time outside

b. ___ to be physically active in your job

c. ___ to do something that makes the world a better place

d. ___ to be in charge (vs. being told what to do)

e. ___ to travel lots

f. ___ to move around a lot from place to place or issue to issue

g. ___ to settle into a place or issue and stick with it

h. ___ to live in a particular place (if this is important to you, what is that place?)

i. ___ job security

j. ___ prestige

k. ___ to be around people (you're a social type)

l. ___ to be left alone

m. ___ to make lots of money

n. __ to have lots of free time to do with as you wish

o. __ to use your mind

p. __ to design, plan, or organize things

q. __ to be working on a bunch of things at the same time (to have many balls in the air at the same time)

r. __ to focus on one thing at a time

s. __ to do something international

t. __ to do something local

u. __ to work at the grassroots level

v. __ to work with power brokers

w. __ to work on projects as a member of a team

x. __ to do something with animals

y. __ to do something with plants

z. __ to work with farmers

aa. __ to work with ranchers

ab. __ to work with environmentalists

ac. __ to work with business people

ad. __ to work on the political front

ae. __ to work with hunters

af. __ to work with fishermen

ag. __ to work with hikers and campers

ah. __ to work with other recreationists (e.g., mountain bikers, horseback riders, ATV users, snowmobilers)

ai. __ to work with loggers

aj. __ to help struggling communities

ak. __ to work for a large governmental organization

al. __ to be around people who see the world as you do

am. __ to be around people who are very different from you

an. __ to promote conservation

ao. __ to preserve natural resources

ap. __ to restore damaged ecosystems

aq. __ social justice

ar. __ to protect endangered or threatened species

as. __ to make polluters accountable for their actions

at. __ wilderness

au. __ to promote the concept of a "working landscape" where people respect but also use the land av. __ to help people get along with one another

aw. __ to raise appreciation for the natural world

ax. __ to fight bad environmental policies

ay. __ to bridge the gap between science and policy making

az. __ to manage or be the steward of a piece of land

ba. __ to teach others about the natural world

bb. __ to have lots of responsibility

bc. __ to lead environmental crusades

bd. __ to get people caring about the political side of environmental issues

be. __ to do naturalist-type activities

bf. __ to have a very well-defined job where there are no day-to-day surprises

bg. __ to work on policy issues

bh. __ to figure out for yourself what needs to be done

bi. __ to be the "number 1 person" (the one who's on the front lines and is the center of attention)

bj. __ to be a support or "number 2" person (not the one who is the center of attention)

bk. __ to write about the environment

bl. __ to give presentations about the environment

bm. __ to design environmental curricula and educational strategies

bn. __ to organize and keep track of things

bo. __ to use your artistic talents

bp. __ to use your musical talents

bq. __ to think about the big picture rather than the day-to-day details

br. __ to always be learning new things

bs. __ to be involved in research

bt. __ to be involved in inventorying or monitoring

bu. __ to be involved in identifying plants or animals

bv. __ to translate technical information into language that lay people can understand and use

bw. __ to use a foreign language

bx. __ to work on projects that have a definite beginning and end

by. __ to work on projects where there's a tangible outcome

bz. __ to be the one who gets things moving

ca. ___ to work with citizen groups

cb. ___ to work for a nonprofit

cc. ___ to be back in school

The quiz above identifies some of the personal interests and values that you hold dear, but some of these are obviously more important to you than others. To rank their relative importance's to you, create a "decision matrix"[3] and make all pairwise comparisons. Begin by counting the number of elements you circled on the quiz and creating a spreadsheet-type matrix that lists, in a row and in a column, the identifying letter of each circled element from before. Now draw in vertical and horizontal lines to create a grid for pairwise comparisons (see p. 155 for guidance) and evaluate each pair of elements. In each pairwise comparison, circle whichever element you favor.

Now total the number of times each element was selected in your priority grid and order the elements from highest to lowest. Beside each element identified as important to you, place the number of times you selected it.

This summary of results from the priority grid provides you with an ordered, explicit articulation of many of the elements that are important to you in a life pursuit. Now, using these ordered elements as brain fodder, work on creating a "farmer's overview" (p. 90) that feels right to you. Work on your farmer's overview until it convincingly describes what you want. You then can use it to describe to others what you're looking for in a life pursuit.

Finding Real-world Opportunities That Excite You

Having met the last objective ("by June 10, to be able to articulate (convincingly and specifically) at least one life pursuit that you'd jump on if you could.)"—you now can move on to finding real-world opportunities that excite you and identify the elements that make them exciting:

Create a folder, entitling it "Promising Opportunities and Exciting Life Pursuits." Over the next couple of weeks, peruse the following[4]:

---

3. If you need a memory jog on how a decision matrix works, see p. 153.

4. If these are not accessible to you, seek assistance from a reference librarian or career counseling center.

- "The Job Seeker—specializing in natural resources and environmental vacancies nationwide." www.thejobseeker.net; tel. (608) 378–4450
- "National Environmental Employment Report—the journal of the environmental careers world." eccinfo@environmentalcareers.com; tel. (757) 727–7895
- *The Back-Door Guide to Short-term Job Adventures,* by Michael Landes. Ten Speed Press; ISBN: 1–58008–669–1

Read about the many opportunities out there, being watchful for opportunities that noticeably excite or intrigue you. When you do find yourself excited about an opportunity, make a hard copy of the description and store it in your folder.[5]

Once your folder holds at least half a dozen descriptions, ask a clear-thinking colleague or friend if s/he would be willing to help you sort through some ideas. Ask the person to scrutinize all of the opportunities in the folder to look for recurring themes, patterns, and tendencies, and to report them back to you. To focus the person's analysis and subsequent reporting, ask this question: "After reading over the opportunities in the folder, what would be a good way to describe my dream-come-true job, opportunity or venture? Please be as specific as you can."

As the person reports back to you, listen carefully and write down what the person has to say. Record the perceptions as bullets; ask the person to be more specific when fuzzy generalizations are offered.[6] If the person says something that surprises you or seems off the mark, ask where the unexpected perception came from. Now ask a different friend or colleague to read over the opportunities in your folder and to offer his or her perceptions, recording these as well.

In what you've recorded, some perceptions probably struck you as being dead-on whereas others left you scratching your head. Use these internal reactions to recast your farmer's overview so that it more accurately captures what's really important to you. Then revise your earlier farmer's overview to better describe what would constitute a great life pursuit for you.

---

5. For now, it doesn't matter if the editions of these publications accessible to you are recent or dated; it also doesn't matter if an exciting opportunity seems out of your league. Remember, the objective you're addressing with this strategy is figuring out what you'd like to do with your life; it's not applying for a job.

6. For example, saying that you seem "interested in nature" doesn't mean much. More specificity is needed.

Wait a few days, then revisit the two sets of bullets as well as your most recent farmer's overview. With a fresh and newly refined understanding of what's really important to you, rework the farmer's overview so that it more accurately captures in print what resides in your soul.

Now What?

By backtracking a bit, you'll see where you've been (implementing the strategy: "finding and dissecting real-world opportunities that excite you and identifying the elements that make them exciting"). You'll also see where you're going—working to meet Objective V: "To craft a detailed plan, by June 30, of what you'd need to do to have a realistic chance of making your desired life pursuit come true."[7]

You met Objective IV when you came up with a lucid farmer's overview; your farmer's overview therefore provides a picture of where you want to go. How to get there? By implementing effective strategies (the "S" in DOC's KEY), of course!

But before rushing off half-cocked, the disciplined thinker in you remembers that there's a "C" in DOC's KEY, and that it precedes the "S." And so, you return to the road map and use techniques from chapter 6 to dutifully seek out constraints and hidden assumptions, listing them in bullet format as you did the perceptions concerning your "dream job." Now you can work on developing strategies for this fifth objective.

After generating lots of ideas, you assess their relative merits (the "K" in DOC's KEY) and decide that the following strategy stands far above the others: *Begin by identifying the exact qualifications needed to make your life pursuit realistic, then see how you measure up to each, and then seek out ways to strengthen whichever qualifications fall short.*

Let's say that you try out the first part of this strategy (the "E" and Y" in DOC's KEY) and identify seven essential qualifications:

1. being knowledgeable about different logging practices

2. having experience working with landowners

3. being a good public speaker

4. knowing how to write winning proposals

5. being knowledgeable about bird habitats

---

7. By working sequentially through DOC's KEY, you've already met the first four objectives.

6. being a competent GIS (Geographic Information Systems) user

7. being able to speak and understand Spanish

With a listing of needed qualifications on paper, you now can assess how you measure up to each. For example, let's say that your self-assessment conclusions are that:

1. You're sufficiently knowledgeable about different logging practices (no problem there).

2. You've worked with landowners in past jobs (no problem).

3. Speaking in public is not a strength of yours (it's something you need to work on).

4. You don't know much about proposal writing (you need to work on this one).

5. You know what you need to know (not a problem).

6. You know some of the basics but that's about it (you need to bolster your skills).

7. You can understand a little Spanish but your speaking skills are practically nil (you have a long way to go).

The self-assessment revealed four situations that are "problems" (3, 4, 6, and 7). Having identified and defined these four problems, you now only need state your goal for each before moving on to the "O" (objectives) and "C" (constraints) for each. These first three steps in DOC's KEY prepare you for "strategy seeking" (the "S" in DOC's KEY). If effective strategies are selected, test run, and implemented successfully, you'll take care of your deficiencies and become adequately prepared for opportunities that most excite you. That's a major accomplishment! Congratulations!

In Conclusion

Before moving on to the final chapter in this book, take a moment and complete your own personal self-assessment. Begin by revisiting your most recent farmer's overview and reading it over carefully. Search out the qualifications needed for you to be adequately prepared for the dream you wish to make a reality. List those qualifications in one column; record the extent to which you meet each qualification in a second column.[8]

The path before you is now clear: find strategies to strengthen your weaknesses and implement those strategies. In so doing, you will prepare yourself for the life you'd like to lead.

8. If you have trouble with this, seek advice from people who are familiar with the type of opportunity that you desire.

Chapter 12

Go Forth and Move the World! Part II

In this last chapter, we'll offer a collection of practical tips for becoming an even more effective environmental problem solver, beginning with a narrative by Dr. Hub Vogelmann, who shares his views on keys to successful environmental problem solving. Dr. Vogelmann's insights and advice are drawn from sixty years of "moving the world" at the highest levels.[1] (To make his points more readable, Dr. Vogelmann uses the term "project" to describe a "problem-solving effort.")

Move the World!

By Dr. Hub Vogelmann

The Doers, the Nondoers and the Undoers
The world has three kinds of people: the doers, the nondoers and the undoers.[2] The doers are the small group of people who get things done while the nondoers sit on the sidelines and watch. The undoers are the naysayers who delight in destroying whatever may be good. The doers are energetic

---

1. In addition to serving as president of the Vermont Nature Conservancy, president of the Conservation and Research Foundation, and special advisor to the Governor, Dr. Vogelmann has served on the board of more than a dozen leading environmental and conservation organizations. Now professor emeritus, Dr. Vogelmann chaired the University of Vermont's Department of Botany for many years and founded the Field Naturalist Graduate Program. Dr. Vogelmann is also world-renowned for his pioneering research on the effects of acid rain on montane ecosystems. By anyone's measure, Hub Vogelmann is a successful world mover.

2. Ignore this part of the "people factor" at your own peril!

and committed; the movers and shakers in the world. The nondoers are merely complacent—either they're satisfied with their lives (and therefore feel no need for change) or they feel powerless (and they sit on the sidelines, maintaining the status quo).

The undoers can, unfortunately, be as energetic and committed as the doers in their determination to keep the world just as it is. Watch out for them, as they have learned how to use innuendoes, distortions, and fabrications to derail worthy projects.

A Powerful Concept

The doers have recognized that they can move the world. They know that with energy and commitment they can make almost anything happen. Their goals must be worthwhile, but, with patience and planning, they can save land for a park or a natural area or help pass needed environmental legislation. In classes I taught at the University of Vermont, I always told my students they can be or do anything they want. They can have a dream and make it happen; one person can have a strong voice in shaping the world. Sometimes, however, a parent, teacher, or mentor must open the door so that a person knows he or she can follow a dream. Here are some of the things to keep in mind.

Focus

First things first. A key element to success is to know exactly what you want to accomplish so that you can state your goals in one or two sentences. If it takes more than one or two sentences, you need to rethink your goals (and objectives) because clear, easy-to-understand statements are key.[3] Don't make the mistake of trying to accomplish too much at once. Well-meaning people often present a detailed, complicated agenda that is often confusing, especially to those not familiar with what they want to accomplish.[4]

A good example of trying to do too much at once is Vermont's bottle bill, which has been held up in the Vermont Legislature for years. A coalition of well-meaning groups has tried to pass a complicated bottle bill that attempts to solve all possible problems[5] dealing with recycling, deposits on

---

3. The "D" and "O" in DOC's KEY: define your problem, goal, and objectives clearly and succinctly!
4. This is why a good farmer's overview is so important.
5. This is the "mega-solution" approach (that always fails).

bottles, cans, fruit juices, wines, and lots more. As expected, confusion has generated opposition to many parts of the bill. A single first step would be to pass legislation that would simply raise the deposit on beer cans and bottles alone, gathering support from groups who want to reduce roadside litter. Since this step also focuses on alcoholic beverages, it would gain other supporters. Forget about the juice cans, fruit mixes, and vegetable cans for now, and pass one small piece of legislation that can be followed with another piece later. That way a stronger bottle bill will eventually emerge to serve all Vermonters well.

Educate and Reach Out

It's also key to educate people and reach out to the public.[6] Plan to be visible but not obnoxious. Use the press and television whenever you can, keeping in mind that local newspapers must publish something every day and television stations must have something for their viewers every night. If your project is worthy, you can reach a large audience through the media. Moreover, there is something about the printed word or mention on television that gives a subject credibility. People pay attention to the local news and they see and hear it again and again. If you are there in print, or on the television screen, your audience is more likely to believe your efforts are worthy. If you need financial support, some of these people will contribute.

Let's assume you have identified a worthwhile project and that you can articulate your goals concisely. That is not enough. You must also market your project effectively. In addition to the press and television, there are other opportunities to reach out to all kinds of people so that they, too, can be made aware of the problem. Plenty of local organizations like the Lions Club, Rotary Club, Optimist Club, garden clubs, church groups, and others all need speakers, so make yourself available.

Some years ago a group of concerned citizens organized to raise public awareness about use and abuse of higher elevations in the Green Mountains. Towers were proliferating on mountains tops, ski trails were being cut on our highest mountain slopes and over-used hiking trails were eroding. Our group, called the Green Mountain Profile Committee, launched an outreach program aimed at informing the public about what was hap-

6. The "people factor" cannot be overemphasized.

pening to our mountains. I spent many evenings traveling around the state with a slide show about the ecology of our mountains and the fragility of the higher elevations.

Early in our campaign we printed hundreds of coloring books, which we distributed free to preschoolers and first grade children.[7] These featured mountains, clouds, water filling our streams and replenishing ground water in the valleys to tell the story of the ecology of our mountains. We called it "Color Me Green." At the supper table parents would ask their children what they learned in school and you can be sure they found out something of the ecology of our mountains. What a great way to reach parents as well as children. When Vermont's Act 250 law was considered in the Legislature (as a strategy to address the problem of overdevelopment at high elevations), there was no opposition to its special protection to the land over 2,500 feet. By that time everyone knew that Vermont's high mountains were fragile and in need of protection.

The Nature Conservancy's experience in saving and acquiring natural areas in Vermont provides many good examples of how education and outreach can be used as a problem-solving strategy. We would enlist the support of a local reporter who would write about our projects. Often the press coverage gave us visibility and promoted a sense of urgency and need, helping us raise money. We also learned that, in fund-raising, many people will contribute $10 or $25. This is all well and good, but we need thousands! We needed to reach an "angel" who would give us all the rest.

I still remember one of the Nature Conservancy's early efforts to purchase Colchester Bog. It had a rich bog flora and was used by biology classes from the University of Vermont. We needed $27,000 to buy it, but after a year of fund-raising from our members, we had collected only $2,000. Then came the angel. A wealthy woman in Pittsburgh heard of our need and sent a check for $25,000! Other fund-raisers tell me this is typical. You receive lots of small contributions but then comes the big one that makes it all happen.

Worthwhile projects may also benefit from a good sales pitch. When the Proctor Maple Research Laboratory at the University of Vermont was destroyed by fire, the director established an "adopt a tree" campaign (strategy) that was enormously successful. Hundreds of people from all over the

---

7. This is an example of a strategy (the "S" in DOC's KEY).

country adopted a tree. Soon a new lab rose from the ashes of the old. The maple trees had their laboratory.

It also helps to know that foundations must, by law, give away 6 percent of their assets each year. Be sure your outreach includes foundations that are likely to be interested in your problem-solving effort. If you receive a foundation grant be sure to send thanks and keep the director informed about your progress. After the project is complete, give more thanks to build a relationship for the future.

Use Psychology

In addition to focusing sharply on your desired outcomes, and educating and reaching out to the public to raise awareness and build support, you'll need to use psychology.[8] Your audience is probably not as familiar with your goal as you are so you should proceed slowly and cautiously. One outspoken person in opposition can destroy or delay your project. Do not be blindsided by an undoer. Avoid controversy, for when it surfaces or polarizes your audience, you may lose ground that's difficult to regain. An essential first step is to learn of your opposition or the potential of opposition and then use psychology to stop it before it starts. Find your worst enemy and talk to him or her before going public. Get this person in on the ground floor. If you try hard enough, there is almost always a common interest or concern. It's also very important to find something you like or admire about your opponent and, then, no matter how hard it is to do, compliment that person. He or she will be flattered, and, noting you have good judgment, will likely look more favorably upon you and your project.

I live in a rural community with open fields, forests, and farmland. When a dairy farm was sold several years ago, a developer purchased the property and planned a sixty-five-house development. It was clearly not in character with our neighborhood, and some community members saw that as a problem. Soon I found myself leading a group of concerned citizens in opposition. The development was turned down by the town planning commission and also by the district environmental commission. The developer greatly reduced the size of the development but he still wanted to build on prime agricultural land. This, in itself, was a problem because prime agricultural land is supposed to be protected and kept in

8. Again, the "people factor"!

agricultural use. Eventually he appealed the case to the Environmental Court. When the trial began, his attorney presented a compelling opening statement to the judge. During a brief break in the proceeding I went to him and told him his presentation was impressive and he was brilliant. It was true. He seemed pleased and during the rest of the trial he was gentle with our witnesses. He could easily have destroyed us but he didn't. We won our case.

It's also important to show respect for everyone—and tolerance—even though at times you may feel insulted or humiliated. People watching a conflict will likely turn against rude and uncivilized behavior. I remember being asked to participate in an acid rain symposium being held at a prestigious university. In my presentation to a large audience, I explained how acid rain was carried eastward by winds that mostly came from the Ohio River Valley. This acid rain was destroying Vermont's Green Mountain forests. The next speaker was an atmospheric physicist who presented slides and tables to show that air trajectories vary in all directions and that my presentation was stupid. It was humiliating. I later learned that the symposium organizers had deliberately set me up for a good fight. I did not give it to them. Instead, I rose from my chair, said "The prevailing winds blow from west to east," and sat down. Later I received compliments for not rising to the bait, and eventually I won my case.

Consider Your Timing

You cannot accomplish a project if you are trying to do it at the wrong time.[9] If the time is not right, hold off until the moment comes. Be patient and it will. The timing call is a test of your good judgment since it involves regional and local politics as well as economic and personal considerations. If you want your town to purchase a park or natural area, for instance, you do not want to push the project when the town has just approved a $2 million school bond and raised taxes.

Sometimes the right time comes when you least expect it. Be prepared to seize that opportunity. Some years ago, before Vermont had strict environmental laws, I often led field trips to the summit of Mt. Mansfield to see arctic plants such as mountain cranberry, alpine bilberry, and Bigelow's sedge, that were left behind when the last glacier melted twelve thousand

---

9. This is an example of a constraint (the "C" in DOC's KEY).

years ago. On one trip I was leading a group of fifteen school teachers. When we reached the summit we were shocked to see the destruction taking place before our eyes.

As the highest peak in Vermont, Mt. Mansfield is the ideal location for radio and television towers, and a construction crew was busy at work. They had bulldozed a road through the thick subalpine forest of spruces and firs. Gravel for the road surface spilled onto the roadside covering the vegetation. A load of unused concrete had been dumped along the roadside. Vehicles had cut ruts into the rare sedge tundra, which had eroded into deep trenches. Empty barrels and junk lay everywhere. We decided right then and there that this was unacceptable and that we needed to do something about it.

A professional writer in the group wrote an emotional letter describing what was happening to the "roof of Vermont." All fifteen teachers signed it. Another teacher took pictures of the destruction. I had the pictures enlarged, made multiple copies, and sent the letter and photos to the governor and to our congressional leaders. I learned later that, the following week, phones in the state capital were ringing off the hook. Shortly thereafter, a crew arrived on Mt. Mansfield to clean up the mess. Later, a mountaintop committee, made up of the users of the mountain, was organized to make policy and to guard the summit from abuse. The group meets on a regular schedule and takes pride in their good stewardship. Today the summit of Mt. Mansfield is regarded as one of Vermont's most precious natural resources and is well protected, thanks to that group who helped make it happen.

On another occasion we were trying to get funding for our research on acid rain. When the *New York Times* carried an article about our work, I immediately contacted a major foundation asking for a large grant. We got it. Later, a foundation board member told me that our timing "was perfect."

Pay Attention to People
Sensitivity to people is essential in all your activities.[10] Do not offend, do not push, respect the views of others and, above all, watch their body language. If you talk with someone about your favorite project, you are soon

---

10. Again, the "people factor"!

aware of how that person feels about it. Body language says it all. From slight changes in the face and eyes you know without a word what this person is thinking. Body language extends to phone conversations as well. Slight changes in voice inflection or pauses are almost as good as eye contact. Take these clues and work with them.

Face to face meetings are most productive in resolving conflicts. A close friend of mine works in a highly prestigious research lab where energy levels and tensions are high. He had conflicts with a coworker about research assignments and sent him daily emails complaining about his work. Each email was more strongly worded than the last. Resentment and anger rose. I asked where this fellow researcher was located and was astounded when my friend replied, "His office is across the hall." I suggested he speak with his coworker face-to-face. They quickly resolved the problems and soon became good colleagues.

Face-to-face meetings are also productive in getting results. When I was chairman of the Botany Department at the University of Vermont, I often had financial problems that needed the attention of the dean of the college. I was always careful to present him the problem and, then, when he looked worried, I would suggest the solution. The dean, like everyone else, is busy, overworked, and does not like to hear about new problems. A solution is more welcome. I clearly remember one day, late in the afternoon, when I had an urgent problem. I called the dean and asked for a short meeting. He was obviously busy and brusquely asked if we could discuss it over the phone. I said, no, but I needed only a few minutes of his time. He asked why we could not handle it over the phone and I replied, "Because I want to watch your eyes." The dean said "O.K., come on over." When I arrived at his office, he was sitting behind his desk wearing sun glasses!

I think I first became conscious of these lessons when I was a college student. There was a beauty pageant in the small town in Ohio where I attended school. There were coeds who dreamed of being the local queen. I remember how one of the contestants, not the most beautiful, watched the body language of the judges. She noticed that their attention was focused on the next contestant and not the one performing on stage. They were watching how the contestant stood up and how she approached the stage. Taking the clue, she rose gracefully from her chair and walked to the stage with style. She was crowned the queen.

Schmoozing

You will be surprised at where and when you will get support for whatever it is you want to do. In everyday conversations you will have opportunities to promote your cause. Do it in a casual manner and do not push. A wandering discussion can often be turned to your project and can lead to pleasant surprises. It is a small world. People like to talk and share ideas. Be ready to pounce! At a party this schmoozing often leads to good results.

Once I was riding on a bus that was taking a group from the Nature Conservancy on a field trip. I struck up a friendly conversation with a pleasant woman seated beside me. Our topics ranged from one thing to another. When I told her we wanted to start a new graduate program at the University of Vermont to train field naturalists, she was clearly interested and wanted more information. Later she sent a check for $5,000 and has sent a similar check every year for the last ten years.

Move the World

Do you ever wonder what makes leaders in business or in academia or in government? For some it could be they had the opportunity because they were in the right place at the right time. For others it could be they had family connections or great wealth. But you can be sure that the best of them arose when they discovered that they could move the world. Focus on your goals, pay close attention to the people around you, and you'll be on your way to changing the world.

Wise Words from Other World-Movers

Below are some other problem-solving suggestions that were learned through the school of hard knocks.
- Spend much more time than you think you should defining the problem and setting objectives. Time spent early on saves lots of time later on.
- Be as specific as you can when you define the problem, but don't be unnecessarily confining. Unnecessary constraints limit your options for solutions.
- Remember: preventing environmental problems is easier and cheaper than solving them once they've developed.

- Fight the urge to take shortcuts to "save time." Wrong shortcuts will cost you dearly later on.
- Carefully document all problem-solving steps on paper. Date everything.
- Summarize all pertinent discussions and telephone conversations on paper so that you have a record of them. Record the date, the time, and who participated.
- Remember that there's rarely one "right" solution to an environmental problem, even if passionate stakeholders believe otherwise. Adjust your expectations so that you focus on the *process* of solving problems rather than on finding the "right" answer.
- Remember to "model the ideal and work backward": decide what outcome needs to become manifest for you to feel that the problem has been solved. (This becomes the goal you'll work toward.)
- Whenever possible, create simplified versions of your problem and solve those simplified versions first. You then can expand your solutions to fit your more complicated problem.
- Visualize problems with sketches, diagrams, and models whenever you can.
- Whenever possible, experiment with solutions. You can learn lots through structured trial and error.
- Don't fall in love with a strategy, you'll lose your ability to evaluate its merits.
- Don't fall victim to, "that's the way it goes" or "it's a fact of life." Faulty assumptions such as these should send chills down your problem-solving spine!
- Environmental decision making is strongly influenced by human values, but values change. Find out—and respect—where others are coming from. It will pay dividends later on.
- Know the time of day when your mind works best. Use that time for hard thinking and idea generation. Save dull, mindless tasks for when your brain is slow.
- Believe that anything is possible until proven otherwise.
- When you're stuck and have been for some time, take a break and get some physical exercise to freshen your perspective.
- Question conventional ways of doing things.
- Question the obvious. If you don't, you'll labor under unnecessary constraints

- Team spirit is important in group problem-solving. Foster it.
- If at first you don't succeed, try, try again—*but in a different way*

As for DOC'S KEY:

- Don't skip steps.
- Don't take steps out of sequence.
- Don't merge steps.

As for "the people factor":

- All "group" problem-solving techniques have the outward appearance of inclusiveness, cooperation, and respect. That impression is helpful if the inclusion is genuine, but empty, hollow, and dangerous if it's simply an overture to appease the crowds. If you really do not want input from others, or if you have already decided what is going to happen in the end, do not alienate others by pretending to value their input. It is better to be authoritative and honest than to be artificially democratic and dishonest.
- Oftentimes it makes sense to ask people to help you with some parts of a problem-solving effort but not others. When you proceed this way, be sure that every participant fully understands the limited scope of his or her participation. If you are not crystal clear about roles, some participants will harbor false expectations about their roles in the problem-solving effort—a situation that creates difficulties beyond your wildest imagination.

And last but not least,

- Practice, Practice, Practice!—but practice with guidance and purpose. Work through this book again in a few weeks, and then work through it yet again. Effective problem solving is a thought process that is not internalized by reading through this, or any, book a single time.
- Persist and you'll prevail. As Calvin Coolidge once said, "Nothing can take the place of persistence. Talent cannot; nothing is more common than unsuccessful people with talent. Genius cannot; the streets are lined with bright people who achieve nothing. Education cannot; the world is full of educated derelicts. Persistence and determination ultimately determine who wins and who loses."

Bibliography

The "People Factor"

Carnegie, D. 1981. *How to Win Friends and Influence People.* Pocket Books, New York, NY. ISBN: 0-671-02703-4

Fisher, R., and W. Ury. 2004. *Getting to Yes: Negotiating an Agreement Without Giving In.* Random House, New York, NY. ISBN: 1-84413-146-7

Variations of "DOC'S KEY"

Harris, R. A. 2002. *Creative Problem Solving: A Step-by-Step Approach.* Pyrczak Publishing, Los Angeles, CA. ISBN: 1-884585-43-4

Heathcote, I. W. 1997. *Environmental Problem-Solving: A Case Study Approach.* McGraw-Hill, New York, NY. ISBN: 0-07-027686-2

Koberg, D., and J. Bagnall. 2003. *The Universal Traveler: A Soft-Systems Guide to Creativity, Problem-Solving, and the Process of Reaching Goals.* 4th Edition. Crisp Publications, Menlo Park, CA. ISBN: 1-56052-679-3

Weeks, W. 1996. *Beyond the Ark: Tools for an Ecosystem Approach to Conservation.* Island Press, Washington, DC. ISBN:1-55963-392-1

Problem Definition and Goal Setting

Odum, E. 1998. *Ecological Vignettes: Ecological Approaches to Dealing with Human Predicaments.* Harwood Academic Pub., Amsterdam, The Netherlands. ISBN: 90-5702-522-1

VanGundy, A. B., Jr. 1980. *Techniques of Structured Problem-Solving.* Van Nostrand Reinhold, New York, NY. ISBN: 0-442-28847-6

Creative Thinking

de Bono, E. 1999. *Six Thinking Hats.* Little Brown & Co., Boston, MA. ISBN: 0-316- 17831-4

Koberg, D., and J. Bagnall. 2003. *The Universal Traveler: A Soft-Systems Guide to Creativity, Problem-Solving, and the Process of Reaching Goals.* 4th Edition. Crisp Publications, Menlo Park, CA. ISBN: 1-56052-679-3

VanGundy, A. B. 2005. *101 Activities for Teaching Creativity and Problem Solving.* John Wiley, San Francisco, CA. ISBN: 0-7879-7402-1

Von Oech, R. 2001. *A Whack on the Side of the Head.* Fine Communications, New York, NY. ISBN: 1-56731-457-0

Critical Thinking and Analysis

Browne, M. N., and S. M. Keeley. 2007. *Asking the Right Questions: A Guide to Critical Thinking.* Prentice Hall, Upper Saddle River, NJ. ISBN: 0-13-220304-9

Nosich, G. M. 2001. *Learning to Think Things Through: A Guide to Critical Thinking in the Curriculum.* Prentice Hall, Upper Saddle River, NJ. ISBN: 0-13-030486-7

Paul, R., and L. Elder. 2001. *Critical Thinking: Tools for Taking Charge of Your Learning and Your Life.* Prentice Hall, Upper Saddle River, NJ. ISBN: 0-386972-4

Whimby, A., and J. Lochhead. 1999. *Problem Solving and Comprehension: A Short Course in Analytical Reasoning.* Lawrence Erlbaum Associates, Mahwah, NJ. ISBN: 0-8058-3274-2

Decision Making

Kahane, A. 2004. *Solving Tough Problems: An Open way of Talking, Listening, and Creating New Realities.* Berett-Koehler, San Francisco, CA. ISBN: 1-27675-293-3

VanGundy, A. B., Jr. 1980. *Techniques of Structured Problem-Solving.* Van Nostrand Reinhold, New York, NY. ISBN: 0-442-28847-6

Environmental Case Studies and Analyses

Ali, S. H. 2003. *Mining, the Environment, and Indigenous Development Conflicts.* University of Arizona Press, Tucson, AZ. ISBN: 0-8165-2312-6

Bartlett, P. F., and G. W. Chase (eds.). 2004. *Sustainability on Campus: Stories and Strategies for Change.* MIT Press, Cambridge, MA. ISBN: 0-262-52422-8

Costanza, R., and S. Jorgensen. 2002. *Understanding and Solving Environmental*

Problems in the 21ˢᵗ Century: Toward a New, Integrated Hard Problem Science. Elsevier, St Louis, MO. ISBN: 0-08-044111-4

Dagget, D. 1995. *Beyond the Rangeland Conflict.* The Grand Canyon Trust, Flagstaff, AZ. ISBN: 0-87905-654-1

Farley, J., J. D. Erickson, and H. E. Daly. 2005. *Ecological Economics: A Workbook for Problem-Based Learning.* Island Press, Washington, DC. ISBN: 1-55963-313-1

Mayer, J. R. 2001. *Connections in Environmental Science: A Case Study Approach.* McGraw-Hill, New York, NY. ISBN: 0-07-229726-3

National Academy Press. 1986. *Ecological Knowledge and Environmental Problem-Solving: Concepts and Case Studies.* National Academy Press, Washington, DC. ISBN: 0-309-03645-3

Index